DISCARDED

TRUMAN'S TWO-CHINA POLICY

For Andy,
Juliana,
and Matthew

June M. Grasso

TRUMAN'S TWO-CHINA POLICY
1948–1950

An East Gate Book

M. E. Sharpe, Inc.
Armonk, New York/London

East Gate Books are edited by Douglas Merwin
120 Buena Vista Drive, White Plains, New York 10603

Copyright © 1987 by M. E. Sharpe, Inc.

All rights reserved. No part of this book may be reproduced in any form without written permission from the publisher, M. E. Sharpe, Inc., 80 Business Park Drive, Armonk, New York 10504

Available in the United Kingdom and Europe from M. E. Sharpe, Publishers, 3 Henrietta Street, London WC2E 8LU.

Library of Congress Cataloging-in-Publication Data

Grasso, June, 1951–
   Truman's two-China policy.

   Bibliography: p.
   Includes index.
   1. United States—Foreign relations—China.
2. China—Foreign relations—United States.
3. United States—Foreign relations—Taiwan.
4. Taiwan—Foreign relations—United States.
5. United States—Foreign relations—1945–1953.
6. Truman, Harry S., 1884–1972. I. Title.
E183.8.C5G72   1987   327.73051   86-31344
ISBN 0-87332-411-0

Printed in the United States of America

Contents

Preface	vii
1. Postwar Problems in Asia	3
2. The China Aid Act and Taiwan	14
3. The Merchant Mission	45
4. Problems for Americans on the Chinese Mainland	57
5. An Inconsistent Policy	80
6. American Interests in China Are Jeopardized	103
7. Taiwan "Abandoned," Taiwan Preserved	126
8. Confrontation in the United Nations	142
9. Conclusion	164
Notes	175
Bibliography	199
Index	203
About the Author	207

Preface

When I started this study, new documents on U.S. foreign relations during the late 1940s had just been released. The obligatory thirty years had passed, and sensitive government papers were declassified. Another look at this period proved revealing, not only for the questions with which this book deals—the formulation of American policy toward China and Taiwan—but also for what happened behind the scenes. President Harry S. Truman tried to keep his goals for Taiwan secret not only from the Chinese Communists but also from Congress, most of the State Department, and the American people. As a result he was criticized for wavering and showing weakness toward the expansion of communism when, in reality, he was remarkably consistent in his attempt to stop Soviet advances in East Asia. The way he and his closest advisers selectively informed key leaders of decisions, purposely misinformed partisans of Chiang Kai-shek's cause, and kept the public in the dark depicts a story of how policy was shaped exclusively in the White House.

I am indebted to many sources of support for the preparation of this book. A grant was provided by the Harry S. Truman Library, where the bulk of my research was conducted. Its staff was always helpful and especially competent in dealing with freedom of information requests. I would also like to acknowledge the assistance of the staffs of the National Archives, the Ginn Library at the Fletcher School of Law and Diplomacy, and the Butler

Library at Columbia University, all of which provided access to indispensable collections. Many thanks also go to my colleagues who helped with the completion of this manuscript. Lynda Shaffer of Tufts University was an important influence at every stage of development. John Gibson from Tufts provided critical analysis and connected me with his State Department friends who were in China after World War II. John Zawacki from Boston University led me to valuable sources that solved several questions. I would also like to thank Anita O'Brien of M. E. Sharpe who edited the final version. The various drafts were typed by Ginny Grasso and the word processing staff at Bentley College. And, finally, I must thank my wonderful family for enduring all that goes with trying to do too many things at once.

A Note on Romanization of Chinese Names

In an attempt to bring some consistency to the spelling of Chinese words in State Department documents, most Chinese names and places have been romanized in the Chinese phonetic system, pinyin, used by Chinese foreign language publications since 1979. The few exceptions are several familiar names that might be confused if changed. For example, the spelling of Chiang Kai-shek and the names of his family members, such as T. V. Soong, remains the same.

TRUMAN'S TWO-CHINA POLICY

1. Postwar Problems in Asia

After nearly four decades since the establishment of the People's Republic of China (PRC), there still exist two Chinas—the PRC and the Republic of China (ROC) on Taiwan. Americans have connections with both. Official diplomatic ties are with the PRC, but the United States trades with both countries. Although sales of military items to the PRC are restricted, American companies sell millions of dollars of sophisticated weaponry to Taiwan each year. The Chinese government on the mainland has tried unsuccessfully to address this issue. PRC leadership claims Taiwan as a province of China that was lost to Japan in 1895 at the end of the Sino-Japanese War. In the years following World War II, while Taiwan's status was entangled in the development of a Japanese peace treaty, the island became the new home of the Chinese Nationalist government, which had fled from Communist advances on the mainland. The PRC has been unable to retake Taiwan partly because of the ROC's connections with the United States.

Since World War II the confusion over the status of Taiwan has centered on President Harry S. Truman's apparent commitment to the Nationalist government of Chiang Kai-shek and not necessarily to the island. Subsequent administrations have also maintained ties to the Nationalists, giving the impression of continued American support for that government as a bulwark against communism in East Asia. This study examines the nature of this commitment and demonstrates that U.S. policy centered not on maintenance of the Nationalists but on the prevention of Taiwan's fall to the Communists. Indeed, Chiang Kai-shek at one point was considered a potential obstacle to this goal because of his reputation for corruption, inefficiency, and mismanagement. It was the island that was desirable because of its strategic importance to American security in East Asia. Truman's dilemma over the future of Taiwan and U.S. relations with the PRC led him to adopt a two-China policy.

In October 1949 the Communist victory in China shattered the vision the United States and its Western allies had had for postwar Asia in the early stages of World War II. The reality of a Communist China in close proximity to U.S.-occupied Japan and South Korea, newly independent India and the Philippines, and war-torn French Indochina made the spread of communism into Asia appear merely to be another step in the USSR's postwar expansion similar to that into Eastern Europe. The policy emanating from the Kremlin challenged American interests worldwide, and it was difficult for most Americans to distinguish between the goals of Soviet and Chinese communism. As Ernest May has stated, "It seemed an unquestionable truth that the United States had . . . suffered a major setback in the Cold War."[1] On the other hand, the U.S. Department of State and its European counterparts were well aware that the Chinese Nationalist government they supported during and after the war was weak, corrupt, ineffective, and unpopular. By early 1948 it was clearly only a matter of time before the Chinese Communists would come to power. This situation presented a dilemma for the Truman administration: Should the United States continue to aid the doomed Nationalist government, or should the Nationalists be abandoned and deliberations begun with the new government?

From mid-1948 until after the outbreak of the Korean War in June 1950, these questions overshadowed American policy in East Asia. The Truman administration was faced with intense opposition to its Asian policies from a variety of sides: among Republicans in Congress, smarting after Truman's surprise electoral victory in 1948 and preparing for the 1950 congressional elections; within the American public, which viewed the United States as having "lost" China; and even at top governmental levels, where several members challenged the administration's policy. U.S. allies were also not entirely supportive. While the State Department persisted in seeking close cooperation and a common policy, particularly with the British, conflicting economic and strategic interests in East Asia prevented the realization of this goal. The relatively early recognition of the People's Republic of China by the British in January 1950 prompted President Truman to comment that they had "not played squarely" with the Americans.[2]

National interests played a significant role in the Truman administration's policy. Secretary of State Dean Acheson, appointed after George C. Marshall resigned on January 21, 1949, prepared a memorandum outlining American interests in China in late 1949, shortly after the Communist victory on October 1. He listed seven categories: (1) American citizens, (2) private property and investments in China of American citizens, (3) trade, (4) U.S. government property, (5) credits due to the U.S. government by the Chinese government, (6) U.S. Foreign Service establishments, and (7) U.S. influence. Acheson determined that the seventh category, American influence in China, "remains one of [the] most valuable assets in that country." He cited the establishment of educational, business, industrial, and missionary organizations as examples of this phenomenon. In conclusion, Acheson commented:

> China is approximately as large as the United States and has a population three times as great. With this in mind, it is evident . . . that the monetary value of our stake in China is comparatively small. But from the standpoint of our national interests much more important than the dollar value of our trade with China and of the physical assets of American missionary and educational institutions, American business and industrial organizations and American Foreign Service establishments is the importance of these as purveyors of American influence, as

symbols of American interest in the Chinese people and as sources of information. Their value to us in this sense represents the accretion of more than a century of governmental and private endeavor.[3]

This "principal asset" would be lost if the United States did not come to an accommodation with the new Chinese government.

A look at the aid programs designed for China by the Truman administration reveals that there was serious debate as China policy evolved after 1948. Differing opinions existed within the White House, State Department, Congress, and the Nationalist government concerning plans for a China aid program, which forced many compromises to be made before the money allocated ever reached China. The slow development from November 1947 to August 1948 of the three major parts of the China Aid Act—a $125-million grant, aid for civilian-type commodities, and the establishment of a joint commission on rural reconstruction—demonstrates the complexity of the negotiations on this matter. Secretary of State Marshall wanted an aid program that would sufficiently protect U.S. interests in East Asia but would not allow the Americans to get "'sucked in,' since it was obviously the Chinese purpose so to involve the United States."[4] Moreover, problems were not solved after the exchange of notes between the U.S. and Chinese governments for the implementation of the program had been finalized. Dissatisfaction with its goals was continuously expressed in both countries. While the Chinese resisted what they considered to be challenges to their sovereignty, the Americans wanted assurances that the money would be properly spent. But by the end of 1948 the State Department was forced to reappraise the goals of its China policy in the face of sweeping Communist victories north of the Yangzi River, divisions among the Nationalist leadership, and the upcoming expiration of the China Aid Act in April 1949.

There were several avenues open in China. Aid to the Nationalist government along the same lines as the China Aid Act could have been continued; increased economic and military assistance could have been given to anti-Communist groups remaining in

South China; or American aid to the Nationalists could have been cut off, thus hastening what was considered to be their inevitable defeat. The year 1949 proved to be significant in the development of American policy toward China. In August the release of the State Department's study entitled *United States Relations with China*, otherwise known as the China White Paper, broadcast the apparent American abandonment of the Nationalist government and acceptance of the Communist victory. On the other hand, the United States continued to recognize and send aid to the Nationalists, even though the government's seat was being pushed further south by Communist advances and President Chiang Kai-shek had resigned his post in January 1949 to set up his base on Taiwan during the summer, leaving Li Zongren as acting president on the mainland.

It is this apparent contradiction in American policy that has perplexed observers and caused many writers to conclude that the U.S. government prior to the Korean War was withdrawing its support for the Nationalists.[5] Recently released documents suggest, however, that this policy was not necessarily contradictory or so different from American policy after the outbreak of the Korean conflict. Even before Chiang's move to the island, Taiwan was considered a strategically important area, and the United States hoped to keep it separate from a Communist mainland government. Economic and political means were used to help save Taiwan from Communist takeover; this was one reason for the continuation of aid to the Nationalist government after its fate on the mainland seemed sealed. As North China came under Communist control, much of the aid targeted for northern ports was diverted to Taiwan. It was hoped that this might prevent an indigenous Communist movement from succeeding on the island, but it would not stop a military advance from the mainland. Before the outbreak of hostilities in Korea, the State Department drew the line at using military measures to secure Taiwan, for several reasons. American military strength was spread worldwide with a concentration in Europe. Budgetary limitations and global commitments made it difficult for the Joint Chiefs of Staff to recommend funding for yet another obligation.[6] It was undesirable to provide additional fuel for Soviet and Chinese Communist assertions that the United States was attempting to wrest the island province from China, and the State Department was preparing for the possibility of recognizing the new Chinese government on the mainland in order to protect American interests there.

The question of recognition of the People's Republic of China after its founding on October 1, 1949, may be examined in detail, but the resulting image is one that reveals no clear policy. Rather, there were conflicts concerning events in China, relations with allies, and input from various sectors of the American government. With the outbreak of the Korean War, the problem was not solved until relations with the Communist Chinese government were normalized in December 1978. Secretary of State Acheson could have been referring to this issue in 1954 when he described the problems he and the president had had in reaching decisions on foreign policy questions. He commented, "In these [difficult] cases the mind tends to remain suspended between alternatives and to seek escape by postponing the issue. There are always persuasive advocates of opposing courses. 'On the one hand' balances 'on the other,' the problem itself becomes the enemy."[7] There were many such problems in Sino-American relations from 1948 to 1950.

Perhaps the most pressing conflict, and the one that Acheson says prevented recognition, was the arrest and detention of the American consulate staff in Mukden by Chinese Communist authorities from November 1948 to December 1949. The failure of continual attempts by American diplomatic personnel to resolve this situation made it impossible for the Truman administration to consider recognizing the Chinese Communists.[8] The Americans suffered mistreatment and were tried in November 1949 for allegedly beating Chinese employees, but U.S. allies, in particular the British, nonetheless announced plans to recognize the PRC. This was a blow to Truman's and Acheson's attempts to formulate a joint Western policy toward the new Chinese government. The

United States seemed isolated in its stand, and yet the president felt there was nothing more to do until Americans were "treated respectfully" in China.[9] The decision on recognition was postponed and Chinese Communist authorities grew impatient. After Communist authorities confiscated part of the American consular compound in Beijing on January 13, 1950, a move which Ambassador O. Edmund Clubb assessed as one designed to hasten recognition, the State Department withdrew its personnel from China.[10]

There were additional sources of tension that further complicated the issue of recognition. Continued American aid to the Nationalists, firmly established on Taiwan by December 1949, offended both the Chinese Communists and the British. Top-level Communist authorities refused to negotiate with American diplomats because of what seemed to be duplicity in American goals. On the one hand, the Americans sought improved relations with the Communists on the mainland, but on the other, they continued to recognize and support the Nationalist group on Taiwan. Moreover, when the Nationalists began bombing congested, poverty-ridden mainland coastal cities in February 1950 with American-made equipment, the United States was blamed for what the Chinese Communist press referred to as "imperialist murder." According to the Communists, there was no point in talking of improving relations until the U.S. government respected Chinese sovereignty over Taiwan.[11] This was something that the Truman administration would not do.

The British were also upset by raids on the mainland since the Nationalists were bombing British-owned buildings and ships and were posing a serious threat to their colony, Hong Kong. In retaliation, the British government authorized its navy to protect its trade and property from Nationalist attack, and several skirmishes took place. Anglo-American relations became strained to the point where the Central Intelligence Agency warned that the deep disagreement over this situation might spill over into other negotiations between the two allies. Acheson's meetings with British officials on this problem led to no solutions.[12]

Yet another feature of American concern over the recognition question involved American economic interests jeopardized by the failure to normalize relations with the new Chinese authorities. Although American trade and investment in China at this time were relatively insignificant and the monetary loss involved in a trade stoppage would do little harm to the U.S. economy, the State Department allowed American companies to maintain economic relations with the Chinese Communists until late 1950, after the outbreak of the Korean War and the withdrawal of American diplomatic staffs. The American-owned Shanghai Power Company, for example, allowed American personnel to remain in China until the Communists took over that plant in December 1950.[13] This seemed to be part of a plan to "keep a foot in the door" and eventually to grant recognition to the new government, although no date for this was set because of the situation in Mukden and other incidents involving Americans in China. Further, the National Security Council had advised the president in February 1949 that continuation of trade would serve to keep the Chinese Communists from becoming totally dependent on the Soviet Union, and mainland China was thought to be a good source of trade for U.S.-occupied Japan.[14] Finally, Anglo-American tension over British relations with the Communists drove Acheson to feel that any trade the Americans gave up would not hurt the Communists but merely go to the British or other Europeans.

The question of recognition also moved into the United Nations, where the issue of Chinese representation was hotly debated after October 1949. The significance of this conflict lay in the fact that China is one of the five permanent members of the Security Council and, as such, has the ability to veto questions coming before that body. China, in theory, held a position equal to that of the other permanent members—the United States, the Soviet Union, Britain, and France. To have the People's Republic of China representing China in the United Nations, it was feared, would give additional support to Soviet positions in the international organization. After the founding of the PRC, it was the

American position to keep the Chinese Communists out of the UN until the United States recognized that government. This was accomplished by convincing American allies to vote against or abstain from voting for the PRC on the representation question.

But this solution proved difficult to maintain. Soviet pressure on the United Nations, including their refusal to participate in debates where the Nationalist representative was seated, challenged the organization's ability to function effectively. This was of grave concern to Secretary General Trygve Lie who, in turn, worked at seating the Communist Chinese representatives. By early 1950, after twenty-four nations had already recognized the PRC, the American position on the Chinese representation question became increasingly unpopular within the United Nations, and the State Department began to show signs of decreasing its pressure on the allies to vote against seating the Communists. But this changed with the outbreak of hostilities in Korea, because the importance of the Nationalists' presence on the Security Council became obvious to the Americans during the first votes on resolutions concerning UN action in Korea. The Nationalists were valuable allies since they consistently supported American positions. The American success at keeping the Communists from replacing the Nationalists shows that on this issue the United States was in control within the international organization.

Another question concerns the significance of the outbreak of the Korean War on June 25, 1950, as a pivotal point in Chinese-American relations. What was the connection between the Korean War and recognition of the PRC? It is unreasonable to conclude that a civil war to unite the Korean peninsula, divided along the 38th parallel by the victorious allies after World War II, would have had such a profound effect on the American attitude toward the PRC unless the conflict was seen as part of a larger Communist threat in Asia. The Truman administration reacted to the North Korean advance by providing military support for the South and placing the U.S. Seventh Fleet between Taiwan and the Chinese mainland. This "neutralization" of the Taiwan Straits was to prevent the Communists, who had been mobilizing along the mainland coast across from Taiwan, from advancing and to stop Nationalist air raids on the mainland. Truman also added that the "determination of the future status" of the island would await the cessation of hostilities in the Pacific, a peace treaty for Japan, or a decision by the United Nations.[15] Given the Chinese Communists' consistent sensitivity over the issue of Taiwan, this statement was perceived as a move designed to delay the recognition process even further.

Not only did improvement in Sino-American relations seem unlikely in the near future, but the continuous setbacks in the American position in China coincided with Soviet advances. Although the State Department had interpreted official Chinese Communist policies after October 1949 as manifestations of Sino-Soviet cooperation, it was also believed that there were some groups in China who reacted negatively to Soviet control. A November 1949 State Department *Weekly Review*, for example, cited evidence of dissension among some Chinese over the establishment of "a Soviet-sponsored [agricultural] program to secure eventual political domination in China." According to the report, this indicated that "the Soviet course in China will not afford plain sailing and may be beset by obstacles presenting opportunities for U.S. exploitation."[16]

In January 1950 a memorandum prepared by Counselor to the Secretary of State George F. Kennan and used by Acheson to summarize American policy to Congress later that month pointed out the areas of Europe and Asia where the U.S. position was in conflict with Soviet communism. The Asian areas that needed continued American support to keep out Communist infiltration included Indonesia, Japan, Korea, Indochina, and Burma. This report considered that the situation in China concerning Soviet control of the new government was "still unclear," and it recommended that recognition be pursued slowly.[17]

The policy toward China began to harden in February when the Soviets and Chinese Communists signed a thirty-year Treaty of Friendship, Alliance, and Mutual Aid. The State Department's Office of Far Eastern Affairs reacted by commenting that the "critical objective" of the USSR in Asia was to consolidate its control over China.[18] In April American consulates in China were closed, and by May, reports coming out of China confirmed the influx of Soviet advisers into the country. Walter P. McConaughy, consul general at Shanghai, related to the State Department at the time of his departure near the end of April that the number of Soviets there forced him to conclude that "the trend in China was completely unpromising and . . . would deteriorate further."[19]

Thus, by June the failure of the United States to come to an accommodation with the Communists and the apparent increase in Soviet influence in China made the outbreak of hostilities in Korea seem to be part of a larger Communist advance. It posed a potential threat not only to South Korea, but also to Taiwan, Japan, Southeast Asia, the Philippines, and perhaps elsewhere throughout the world. The military action taken by the United States to protect these areas prevented further Chinese-American negotiations concerning recognition. With the entrance of the Chinese Communists into the Korean War in November 1950, relations between the two countries did not improve for nearly thirty years. The recognition of the Nationalists on Taiwan as the Republic of China provided the basis for Truman's two-China policy.

2. The China Aid Act and Taiwan

One of the main components of postwar aid to China was Title IV of the Foreign Assistance Act of April 1948, the China Aid Act, which authorized a sum of $463 million to the Chinese Nationalist government. Revenue from this program eventually was sent to Taiwan and initiated the American commitment to the island. On the one hand, this aid appears to have been part of a continuing policy of military and economic assistance to China that began with Japan's declaration of war on that country in 1937. More than three billion dollars had been sent to help the Chinese government fight the Japanese and recover after the war.[1] On the other hand, the bill sparked considerable controversy in the United States. The China Aid Act was perceived differently by various Americans concerned with the U.S. role in China. One congressman, for example, said that it gave the Nationalists "at long last some hope,"[2] but a State Department official declared that he and his colleagues were "faced with the likelihood that all of this equipment that came from this . . . would sooner or later fall into the hands of the Chinese Communists."[3]

Several writers have argued that it was the overwhelming power of a group of Nationalist supporters in Congress known as the China bloc that forced this bill on a reluctant Truman administration.[4] The complexity of the negotiations suggests, however, that this was not the case. Although domestic pressures were a factor, there was also the variety of opinions concerning plans for China within the White House, State Department, Congress, and the Nationalist government. Some compromise took place, but in the end, policy was formulated by the president and his secretary of state. Their goal was to protect American interests in East Asia but not allow, according to Secretary of State George C. Marshall, the Americans to get "sucked in."[5] The formulation and implementation of the China Aid Act were reflections of Truman's and Marshall's attitudes.

An Aid Program Is Developed

The initial move in presenting an aid program for China to Congress was made by Secretary of State George C. Marshall at a Joint Session of the Senate Committee on Foreign Relations and the House Committee on Foreign Affairs on November 10, 1947. Marshall stated:

> The situation in China continues to cause us deep concern. The civil war has spread and increased in intensity. The Chinese Communists by force of arms seek control of wide areas of China.
>
> The United States and all other world powers recognize the national government as the sole legal government of China. Only the government and people of China can solve their fundamental problems and regain for China its rightful role as a major stabilizing influence in the Far East.
>
> Nevertheless we can be of help and, in light of our long and uninterrupted record of friendship and international cooperation with China, we should extend to the government and its people certain economic aid and assistance. A definite proposal is under preparation for early submission.[6]

In response to a question by Senator Arthur H. Vandenberg (R., Michigan), chair of the Senate Foreign Relations Committee, Marshall told the committee that the administration would ask for $300 million for economic aid to China for a fifteen-month period.[7]

W. Walton Butterworth, director of the State Department's

Office of Far Eastern Affairs, claimed that the decision to develop an aid program came after several months of study by the State Department and the working-level group in the National Advisory Council concerning the Chinese government's balance-of-payments deficits and the rapid economic deterioration threatening the country. During the last half of October 1947, the State Department formulated a plan designed to cover China's need for essential imports, also referred to as balance-of-payments aid, to retard the development of severe economic damage. This prompted Marshall's November 10 statement, and subsequently a program evolved within the department.[8]

On November 28, Melville H. Walker, assistant chief of the Division of Investment and Economic Development, prepared a memorandum on economic aid to China that revealed the State Department's preliminary agreement concerning the character and magnitude of the program. It called for assistance totaling $400 million for the period from April 1, 1948, to June 30, 1949. Of the total, $300 million would be allocated for balance-of-payments aid, and the remainder would pay for reconstruction projects. The program was designed specifically to exclude any provision for direct military aid and currency stabilization.[9]

This proposal was changed by December 30, when Clinton T. Wood, deputy to the assistant secretary of state for economic affairs, announced that the funds would be increased to $485 million, of which $435 million would be allocated for essential imports and the remainder for reconstruction projects. The period covered by the program was advanced to eighteen months, beginning January 1, 1948. According to Wood, developments in China's financial situation had resulted in a more rapid disappearance of financial resources than had been anticipated.[10]

The debate on the China Aid Bill came at a time when both the military and economic situations in China lent an air of urgency to the need for aid. There had been a change in the Chinese military picture during the summer of 1947, when Communist forces went on the offensive. Their southward advance put them in control of much of the North China plain and its communications and transportation systems, crippling the Nationalists there. Despatches from American consulates in Nanjing, Shanghai, and other cities were consistent in their agreement on the territorial advances and military victories by the Chinese Communists and predictions for greater Nationalist losses. Economic conditions, they reported, were deplorable. The consul general at Shanghai, John M. Cabot, described the situation in that city in early February 1948. He "was struck by the perfect field for the spread of Communism which [Shanghai] afforded." The bulk of the city's population seemed on the "ragged edge of subsistence." He commented, "Too many of the wealthy people have made their money by means which are, to say the least, devious, and I am afraid spend it in ways which are, to say the least, heartless." Any middle or intellectual class was being squeezed out by severe inflation.[11]

The reports from Nanjing, the Nationalists' capital, were equally disheartening. Ambassador John L. Stuart wrote on January 29 that the embassy "cannot stress too strongly [the] fact [that the Chinese] Government's military situation is critical and that [the] Government is completely unable to regain lost ground or even maintain [a] hold on areas under its control."[12] Stuart was also concerned about the government's financial affairs. On March 13 he cabled Marshall that an official of the Central Bank of China had confidentially informed him that the bank's U.S. dollar reserves were "virtually exhausted" and only "some" gold and silver bullion reserves remained. Stuart thought that this report might be true because of a sharp break in the black-market Chinese national currency rate and Nationalist requests for advances of American funds for petroleum and cotton purchases.[13] As a result of this situation, Butterworth was dissatisfied with the department proposal as outlined by Wood because of "its probable inadequacy to accomplish the minimum objective of the program—to prevent further rapid deterioration in China." He also claimed that Congress would be inclined to accept the State Department's assurance that such a proposal would be sufficient to stop China's economic deterioration. And, if it failed, "responsibility will be laid at the Secretary's [of State] door."[14]

Butterworth was apparently concerned about the reaction of pro-Nationalist American public opinion, particularly Chiang Kai-shek's supporters in Congress, known as the China bloc. Nationalist sympathizers such as Senator Styles Bridges (R., New Hampshire) and Representative Walter H. Judd (R., Minnesota) had been pressuring the State Department for increased aid to China.[15] Butterworth noted that they would be concerned that the Chinese government might purchase munitions and banknotes with the money from this program and exhaust the government's foreign exchange assets.[16] The efforts of the China bloc were further aided by the presence of a Chinese technical mission, which arrived in Washington, D.C., on January 29 to take part in the formulation of the program. The Nationalists desired a four-year plan totaling $1.5 billion. They wanted $500 million for each of the first two years, $300 million for the third year, and $200 million for the fourth year.[17]

Requests by the China bloc and the Chinese technical mission for even more money than had been suggested in the December 30 proposal went unheeded. Admiral Sidney Souers, executive secretary of the National Security Council, also pushed for a stronger aid plan. He has explained why pressure from the China bloc and other Nationalist supporters had minimal impact on the formulation of the China Aid Act. He spent several hours trying to persuade Secretary of State Marshall to commit American support to the Nationalists but was rebuffed because of Marshall's disenchantment with Chiang Kai-shek. Souers commented that "Marshall gave me a long dissertation covering his arguments about Chiang's policy, and said we couldn't win. . . . We would go down to defeat with him if we give our wealth and troops to his cause. He said Chiang wouldn't listen to advice."[18]

Copies of the tentative program were also sent to the U.S. embassy in China in January. Upon receiving the information, Ambassador Stuart reacted by criticizing the department's goal of sending aid for essential imports and reconstruction projects while letting the Nationalists take full responsibility for the course of the civil war. He and his colleagues were perturbed by the implications of the proposal and felt that they had failed to create in the department a realization of the problems that confronted American policy in China. These involved the failure on the part of virtually all spheres of the Chinese government studied to effect improvements due to blunders, mismanagement, and misfeasance. As a result, Stuart and his colleagues believed that an aid program could not be effective if activated through the Chinese government unless the United States developed requisite plans and supervised their implementation and execution on an advisory basis.[19]

Stuart concluded by admitting that these suggestions involved responsibility on the part of the U.S. government. Without the assumption of such responsibility, it was difficult to see how the situation in China could be restored in favor of the United States. He warned that there had been rumors indicating that certain elements within the Nationalist party were considering Soviet mediation between the Communists and the Nationalist government.[20]

Of all the pressures placed on the Truman administration, the threat of Soviet involvement in the Chinese civil war should have been a powerful weapon for Nationalist sympathizers. One year earlier the Truman Doctrine had outlined an American commitment to halt the spread of Soviet-inspired communism in Europe through massive aid and the sending of advisers to Greece. The Nationalists were hoping for a similar commitment to China. But the suggestions and warnings of Butterworth, Souers, the Chinese technical mission, the China bloc, and the American embassy in China had little impact on the State Department's program. The proposal was formally presented to Congress on February 18, accompanied by a message from President Truman. It provided for $570 million in aid until June 20, 1949. The program included financing, through loans or grants, essential civilian-type imports to China in the amount of $510 million, and $60 million for reconstruction projects initiated before June 30, 1949. In his message the president said, ''We can assist in retarding the current economic deterioration and thus give the Chinese government a further opportunity to initiate the measures necessary to the establishment of more stable economic conditions. But it is, and has been, clear that only the Chinese government itself can undertake the vital measures necessary to provide the framework within which efforts toward peace and true economic recovery may be effective."[21]

When the American embassy received its copy of the plan, Marshall addressed Stuart's earlier criticisms of the program by directing the ambassador to avoid any implication that the U.S. government was or should be assuming responsibility for or underwriting China's economic recovery or the Chinese government's military effort. He was also to avoid enlarging on expected results and to "remember that [the] proposed program is not [a] complete or long-range economic recovery program, [it] is not expected to stabilize currency, end inflation or provide for large-scale reconstruction."[22]

The China Aid Bill Comes Before Congress

The next step was to assure the bill's passage through Congress. At this stage the plan became linked with aid programs for Europe, also before Congress, because Nationalist sympathizers, particularly those in the House Foreign Affairs Committee, were able to exert additional pressure so that the bill conformed more closely to European programs. But the influence of what proved to be a vociferous minority was minimal.

On the morning of February 19, Charles E. Bohlen, counselor to the State Department, met with Senator Arthur Vandenberg, chair of the Senate Foreign Relations Committee, to discuss the various aid programs to be considered by Congress and devise a plan to expedite their passage. The significance of such a meeting is perhaps not apparent until one understands the close relationship between Vandenberg and Under Secretary of State Robert Lovett, Bohlen's superior. Dean Acheson, also an under secretary of state in 1948, has written that Vandenberg's collaboration with the administration, coupled with his "immense" influence among colleagues of both parties, was a valuable aid in obtaining support for the administration's proposals. Acheson refers to Vandenberg as "a master of maneuver and a superb advocate."[23] Bohlen informed Vandenberg of a decision by the House Foreign Affairs Committee to start hearings on China on February 20. Vandenberg said that he thought it was a good idea for the House to start the hearings because his committee would not be able to hear testimony concerning foreign affairs until the European Recovery Program passed the Senate, approximately March 25.[24] Then Vandenberg outlined a schedule that he felt would avoid Senate debate on the proposed aid programs—the China, Greek-Turkish, and Trieste aid bills. The timing and relationship of the three bills to the European Recovery Program was important for expeditious passage of the State Department's proposals. The best schedule would be:

(1) That the Senate would concentrate on ERP and would not have the China, Greek-Turkish, or Trieste bills introduced in the Senate.

(2) The Senate Foreign Relations Committee would have hearings on these measures just as soon as ERP had been passed by the Senate.

(3) That the House would proceed with its schedule to terminate hearings on all these foreign aid measures by March 2 and then get all passed by the House as soon as possible which he thought could be done by April 1; that is ERP with the China program as part 2 of the same bill in view of the fact that the [Economic Cooperation] Administrator was to handle both, and the Greek-Turkish and Trieste as separate bills.

(4) In the conference between the two Houses which would follow immediately upon the passage of these measures by the House, the Senate would only have their ERP bill and would adopt with such amendments they saw fit the House bills on China, Greece-Turkey, and Trieste.[25]

Also discussed was a comparison of the China Aid Bill with the Foreign Aid Act, the latter dealing with European recovery. The two bills were similar in many respects except for specific references to amounts of money, types of commodities, and conditions relating only to China. But there were also several important differences. Section 2 of the Foreign Aid Act, which was left out of the China Aid Bill, contained a reference to alleviating conditions of hunger and cold as well as preventing serious retrogression. On the other hand, Section 3(a) of the China bill gave the president certain discretionary powers "in order to guard against eventualities" so that he might furnish aid to China under conditions other than those spelled out in the bill. This authority was not granted for the aid program to Europe and proved to be important later when the Chinese Nationalists moved to Taiwan.[26]

The House Committee on Foreign Affairs began hearings on the proposal on February 20, 1948. Following an executive session with Marshall and a series of public hearings, the committee concluded that military aid should be provided in the China Aid Bill. Representative John M. Vorys (R., Ohio), a member of the committee, lent his support to this proviso and was among those Republicans who were advocates of a combined aid program to Europe and China.[27] They deducted $150 million from the $570 million total proposed by the State Department and added it to Title II of the Foreign Assistance Act of 1948, which dealt with military aid to Greece and Turkey. The result was a provision of $150 million in military assistance to China of the same character as that allocated to the two European countries. An explanatory statement accompanied the proposal:

> China requires aid of the same type as that extended to the European countries under Title I [ECA economic aid] but it also requires aid to the special types included in the program of Greek-Turkish aid extended under the Greek-Turkish Aid Act of 1947.
> The committee, having in mind the weighty testimony received on the need for military aid, separated the amount requested into two parcels, transferring $150 million to provide aid for China under Title III.[28]

The Senate Foreign Relations Committee then considered the proposal, reducing the total amount of funds allocated from $570 million to $363 million and the period to be covered by the Foreign Assistance Act from fifteen to twelve months. The character of the $150 million for military aid proposed by the House Committee was also changed. To avoid having China placed in the same category as Greece and Turkey with respect to military aid (Title III), it was decided that a grant not to exceed $100 million would be allocated on such terms as the president might determine without regard to the restrictions within the ECA.[29]

This new feature, a $100 million grant, was proposed by the Senate Foreign Relations Committee because members were strongly opposed to an extension of military aid to China similar to that for Greece and Turkey because of "the possible involvement and commitments entailed in such action." According to State Department representatives present at the meetings, committee members felt so strongly on this point that the proposal was put forth with the knowledge that past experience clearly indicated the probability of misuse by the Chinese administration of uncontrolled funds of this nature. The committee's report concerning the revised China Aid Bill revealed that it was aware of the probable use for the $100 million grant. "In view of the Chinese requirements for military supplies, it may be assumed that the Chinese government, on its own option and responsibility, would seek this grant for such supplies."[30]

During the Senate debate on the China aid program on March 30, Senator Vandenberg attempted to explain the committee's intentions in his introductory address:

> The Committee on Foreign Relations wishes to make it unmistakably clear, in this, as in all other relief bills, that there is no implication that American aid involves any continuity of obligation beyond specific, current commitments which Congress may see fit to make. . . . We do not—we cannot—underwrite the future. . . . It is a duty to underscore this reservation in the case of China because we find here many imponderables as a result of military, economic and social pressures which have

understandably undermined her stabilities. . . . We cannot deal with the Chinese economy on an over-all basis, as we have done in the European recovery program. China is too big. The problem is too complicated.[31]

Vandenberg then spoke of the $100 million grant. He told the Senate:

Your committee believes, as a matter of elementary prudence, that this process must be completely clear of any implication that we are underwriting the military campaign of the Nationalist Government. No matter what our heart's desire might be, any such implication would be impossible over so vast an area. Therefore, for the sake of clarity, we prefer to leave the initiative, with respect to these particular funds, in the hands of the Nationalist Government.[32]

The Senate debate that followed was dominated by the bill's opponents, who took the opportunity to present a variety of reasons for what turned out to be the minority viewpoint. Senators Wayne Morse (R., Oregon) and Claude Pepper (D., Florida) wanted assurances that the funds designated would reach the common people of China and not the "corrupt" Nationalist government.[33] Pepper was applauded after he stated that "while we are not going to yield to the aggression of Communism, we are not going to embrace as brothers the Fascists or those who have philosophies of government which are not basically and essentially democratic in character." He asked whether it would be possible for the bill to outline "basic essentials of democracy" that must be met before aid was given.[34]

Of the many other expressions of opposition that characterized the testimony, one that was supported by several senators was the opinion that the China Aid Bill appeared to be a prelude to American involvement in another war. Senator James P. Kem (R., Missouri), citing a statement by Franklin Roosevelt in October 1940 that "our boys are not going to be sent into any foreign wars," questioned whether American aid to Greece, Turkey, and

China was another beginning to increasing American involvement in the wars of those countries. After several explanations that the proposed bill concentrated on economic and not military aid, Senator Tom Connally (D., Texas) stated that "anything which supports human beings in the way of food and the like in a measure is military aid, if the people are in the field fighting."[35]

Among the lengthiest arguments in the debate were those made by senators who felt that the United States should keep its money at home. Senator Kem cited segments of a letter by George Washington in 1795, which stressed the need for Americans to stay out of European affairs. Kem argued that this was sound advice for the United States in 1948, and it should pertain to Asia as well as Europe. Of primary concern to Kem and others was the amount of money sent to China. A vocal minority stated that "our first concern should be to build a strong economy at home."[36] Many anecdotes focusing on the economic plight of American citizens were told to stress the point that the U.S. government planned to allocate $463 million to China while American GIs, farmers, and welfare recipients, for example, were left with insufficient funds.[37]

After debate in the Senate was concluded, the Senate and House versions of the various aid bills were referred to a conference committee. This produced the Conference Report of April 1, 1948, which placed all the aid bills into one act entitled the Foreign Assistance Act of 1948. The China Aid Bill became Title IV of this act and was entitled the China Aid Act of 1948. The Foreign Assistance Act of 1948 was passed by Congress as reported out of conference. The Senate version of the China Aid Bill was used in the Foreign Assistance Act with the exception of the wording of a paragraph dealing with the creation of a Joint Commission on Rural Reconstruction, taken from the House version.[38] Also, the length of time covered by the program was reduced from eighteen to twelve months, and a change was made in the amounts authorized under the two types of aid. The new totals were $338 million authorized for economic and reconstruction aid and a $125 million grant to be used for whatever purpose the Nationalists wished. Congress passed this revised bill on April 2, and it was signed by the president on the following day.

The passage of the China Aid Act came after seven months of planning within the State Department followed by revision in congressional committees. The initial program presented to Congress was the product of debate within the State Department, which had resulted in little compromise between Secretary of State Marshall and his advisers in the office of Far Eastern Affairs, the National Security Council, and the embassy in China, such as Butterworth, Souers, and Stuart. Souers has commented that Marshall's influence and his feeling that the United States should not commit its military and economic resources to saving the Nationalists can be blamed for the "loss" of China.[39] This was because the China Aid Program was not designed to be analogous to the European aid programs in that it did not commit the United States to a defeat of the Communists.[40] Such an undertaking would have severely taxed American military strength, which was already spread across southeastern Europe and eastern Asia. Marshall formulated the program to be one of economic assistance to retard the deterioration of the Chinese economy.[41]

It is within congressional authority to allocate the funds for such a program, and it was in Congress, particularly in the House Committee on Foreign Affairs, where the China Aid Bill was linked with the European aid proposals and was modified to include a military aid proviso. But this was changed by the Senate Foreign Relations Committee, which expressed a view similar to that of the administration—China should not be placed in the same category as southeastern Europe with respect to military aid. As a result, a grant replaced funds providing specifically for military aid, thus absolving the U.S. government from any commitment to a Nationalist victory while satisfying those in Congress who insisted on giving military aid to the Nationalists.

The Senate Foreign Relations Committee's reluctance to provide military aid to China, and the fact that the European Recovery Program was passed in the Senate before debate on the China Aid Bill began, challenge the conclusion that it was necessary to link China and Europe in order to get congressional approval for a European aid program. Clearly, there were those who supported this linkage, such as China bloc members Vorys, Judd, and Bridges, but there was also another group who balked at sending aid to China, such as Pepper, Kem, and Morse. Therefore, while Marshall formulated the initial aid proposal, a divided Congress compromised to produce the resulting limited aid bill.

Renewed Requests for Aid to China

By late 1948, the State Department was forced to review its policy toward China. The Communists were consolidating their position north of the Yangzi River, and more victories seemed imminent. The Nationalists were stepping up their efforts to garner support for their cause, yet their fate seemed sealed. The expiration date for the allocation of funds under the China Aid Act was but a few months away, on April 2, 1949, and only a fraction of the funds had been spent by this time. American interests in China, in the event of a probable Nationalist defeat on the mainland, had to be reappraised.

Discussions on several issues that set the stage for American policy during the following years, such as the continuation of aid programs and strategic interests in the area, evolved during the last few months of 1948. In September, talks began on the question of whether to continue aid to China after the China Aid Act's expiration. A variety of proposals were offered by both Americans and Chinese. There was no attempt to separate the issues of economic aid and military assistance as had been done in the China Aid Act; the suggestions reflected the Nationalist government's desperate situation.

One proposal that received considerable attention from supporters of strong American action in China was put forth in October by Roger D. Lapham, chief of the Economic Cooperation Administration (ECA) in China. Lapham's program, which had been drawn up with the assistance of Under Secretary of the Army William H. Draper, General Thomas S. Timberman (chief of the Operations Group, Plans and Operations Division, General Staff, U.S. Army), and Naval Aide to the President Robert L. Dennison, was a combined economic and military plan for assistance similar to the Greek aid program established as part of the Foreign Assistance Act of April 1948.[42]

Chinese Nationalist leaders gave full support to these recommendations, which implied the use of American military advisers and the possible deployment of American military forces to reinforce Nationalist armies. They had been pushing for such a proposal since the China Aid Act was in its planning stages. Hollington Tong of the Chinese Government Information Office made a lengthy statement praising Lapham's proposal, calling it "a heartening augury for China." He claimed that such a program "would give virtual assurance of China's triumph over its present difficulties."[43] These sentiments were also heard within the U.S. government, particularly from members of the China bloc, who had favored a comprehensive military aid program for China at the time of the China Aid Act's passage.

In light of the apparent support for this program, Acting Secretary of State Robert A. Lovett sought an estimate of the costs of the military assistance program envisioned by Lapham "for the purposes of planning." In a memorandum to Secretary of Defense James Forrestal, he requested a cost analysis of two types of programs: (1) a plan to provide for the replacement and maintenance of American ground, naval, and air equipment in Chinese possession and equipment furnished from the $125 million grant, to furnish ammunition for American weapons in the possession of Chinese forces, and to provide aviation gas to enable the Chinese to meet combat needs; and (2) an "all-out military aid program" which included U.S. military advisers on a scale similar to those in Greece, to provide the Chinese government with the facilities necessary for stopping the Chinese Communist advance, and eventually to destroy organized Communist strength.[44]

Forrestal, who had been among the advocates of a stronger military aid program to the Nationalists, never responded to Lovett's inquiries. While Lovett admitted in the memorandum's conclusion that an accurate prediction of the program's needs would require a lengthy and detailed study, he merely wanted "rough estimates" from the military establishment. Lovett and other State Department officials, who had consistently resisted the idea of sending military aid to China, apparently realized that the rapidly changing situation there made it impossible for any such program to be seriously considered.[45]

The Nationalists also came forth with a series of requests for aid through their embassy and other special representatives. One effect of these requests was to garner publicity for the Nationalists' plight.[46] On the other hand, they seemed to have little impact on those who determined policy. On November 9, Ambassador V. K. Wellington Koo delivered a message from Chinese President Chiang Kai-shek reporting that Communist forces were within striking distance of Shanghai and Nanjing and seeking increased military aid as well as the participation of American military advisers to direct military operations. Chiang added that he would be "most happy to receive . . . as soon as possible a high-ranking military officer who will work out . . . a concrete scheme of military assistance."[47] Within a few days Jiang Tingfu, Chinese representative at the United Nations, delivered another document to Secretary of State Marshall outlining the Chinese Foreign Office's suggestions for a new aid program. The request was for a plan on par with the four-year European Recovery Program, including a three-year program of economic aid consisting of an annual allocation of $450 million for economic assistance and $550 million in military aid, and assistance in technical services.[48]

Truman replied to both of these messages on November 12 by reiterating the status of American aid to China at that time. Every effort was being made to expedite shipment of military and economic assistance to China under the provisions of the China Aid Act. He also mentioned that Major General David G. Barr, director of the Joint United States Military Advisory Group in China, was there to offer Chiang advice. Truman made no mention of a renewed aid proposal or any changes in the program already authorized.[49]

In December the American ambassador in London, Lewis W. Douglas, received word that the Chinese ambassador in London, Zheng Tianxi, was pushing the British to use their influence with the United States to have more aid sent to China. The British Foreign Office's reply had been that no amount of aid to China could alter the situation at that time since the "rot" in the Nationalist government was "too deep."[50]

Of all the requests by the Nationalist government for assistance, the one on which the most attention was focused was the trip to the United States by Madame Chiang Kai-shek to appeal for help. Madame Chiang arrived in Washington in December with the goal of getting large-scale economic and military aid for the Nationalist government. Her visit predictably received much publicity; she was depicted as an American-educated, English-speaking, Christian spokesperson for a regime dedicated to preserving "democratic" principles in China.[51] Members of the State Department and the American embassy in China were unhappy with her decision to make the trip because it was "doomed in advance to failure" and would only serve to give additional fuel to Nationalist sympathizers in the United States.[52] Her visit was discussed at a cabinet meeting on November 26, where the president raised the question of her plans. Secretary of State Marshall explained that if she were allowed to come, she would talk to the public and the press. This would put the United States government in a position of having to refuse further aid. Marshall stated that "the Nationalist Government of China is on its way out and there is nothing we can do to save it." The questions for the cabinet were: Should they "play along with the existing government and keep the facts [that the Nationalists were losing the civil war] from the American people, or should Madame Chiang be barred from the United States?" The president replied, "We should not bar her from the country." Marshall then recommended that Madame Chiang be allowed full courtesies if she came to the United States.[53]

Madame Chiang met with Marshall on December 3 and outlined three urgent requests: (1) The U.S. government should make a statement in opposition to the Communists in China and in support of Chiang's government; (2) an outstanding American soldier should be sent to China and be the "spark plug" of the Chinese military effort along with American officers to supervise Chinese staff officers and commanders; and (3) greater economic assistance was needed.[54]

The official position toward Madame Chiang's mission was to give no indication of any change in the American position and make no commitments beyond the terms of the China Aid Act.[55] After nearly four weeks of meetings with U.S. government personnel, Madame Chiang tried to pressure the secretary by explaining that she had received a message from her husband earlier in the day stating that members of the Nationalist government were pressing him to make an accord with the Soviets if no more American aid was forthcoming. He would make his decision that day to resign based on the outcome of her meetings and in the event that an agreement with the Soviets was to be made.[56]

Madame Chiang was bothered by Marshall's claim that he had little power to act on her behalf. In a later conversation with Brigadier General Marshall S. Carter, she stated that both Truman and Lovett had made her think that Marshall had "the final word in the matter."[57] Marshall, however, had been hospitalized during December and had not been officiating as secretary of state. He claimed that he had been unaware of events in China for several weeks, and messages she had transmitted through his wife could not be passed on to the State Department without some confusion or omissions.[58]

Madame Chiang then met with Lovett, asking him to make a commitment concerning the issue of aid that would have an impact on her husband's decision to resign. Lovett explained that the United States could not be placed in a position to decide for Chiang whether he should resign. Truman's November 12 reply to Chiang reiterating the terms of the China Aid Act was to be considered the latest statement on American policy toward China. Lovett agreed, however, to discuss with the president the idea of releasing another statement and Madame Chiang's urgent plea for another military adviser to be sent to China.[59]

The Strategic Importance of Taiwan

At about this same time the situation on Taiwan became the focus of discussions within the State Department. Several questions, such as the strategic importance of the island, the possibility of transferring American naval personnel there from the Communist-surrounded Qingdao base in North China, and requests by the Nationalists to divert American military aid to the island were considered. On November 8, 1948, Acting Secretary of State Lovett directed a memorandum to the National Security Council requesting an appraisal by the Joint Chiefs of Staff of the strategic implications for the security of the United States should a Communist administration control Taiwan. The Joint Chiefs expressed the opinion that the strategic implications in such an event would be "seriously unfavorable." Once the mainland had come under Communist control, they indicated, the U.S. government would no longer have access to its harbors, air bases, and coastal railroad terminals in case of war. Therefore, Taiwan would be most valuable as a base for staging troops and controlling air and sea operations. An unfriendly administration on Taiwan would also have the potential to dominate adjacent sea routes, thereby posing a threat to American interests in Japan, the Ryukyus, the Malay peninsula, and the Philippines. Taiwan, moreover, was a major source of food and other materials for Japan. The recommendation was thus made that "it would be most valuable to [U.S.] national security interests if . . . Communist domination of Formosa could be denied by such diplomatic and economic steps as may be appropriate to insure a Formosan administration friendly to the United States."[60]

Shortly after this statement was formulated, the National Security Council discussed the question of whether to transfer U.S. naval forces stationed in Qingdao to Taiwan. In a memorandum dated December 1, 1948, the choice of Taiwan as the site for a naval base was considered. The island was deemed an inappropriate location because of the repercussions that would result from the movement of the U.S. Navy there. The memorandum stated that "the Department of State recognize[s] the strategic importance of Taiwan and [is] fully cognizant of the undesirability of its passing under the control of a Chinese Communist-dominated government," but the stationing of American forces there would cause mainland Chinese, including "predatory politicians and carpetbaggers," to flee to the island in large numbers with the belief that they would be protected by the navy. Furthermore, such a move would only serve to support Communist charges that the U.S. government planned to sever the island from China. This might have the disadvantageous effect of jeopardizing the safety of Americans in China and endangering U.S. interests.[61]

The decision not to establish an American naval base on Taiwan in late 1948, of course, made the goal of keeping Taiwan out of Communist hands more difficult. It was understood in the State Department that although the Joint Chiefs were not prepared to support a military conflict between Americans and Chinese Communists over the island at this time because of American military commitments in Europe and Japan, they had urged that "all possible political and economic means be taken to keep it out of the hands of the Chinese Communists."[62] Several plans, including one to send massive amounts of economic and military aid and to work with Taiwanese independence groups, were formulated during the next few months.

Also during November and December 1948, questions concerning shipments of supplies to China paid for by the China Aid Act were discussed. Diverting much of this aid to Taiwan fit easily into the economic and military aid program because the aid act allowed for goods to be sent to areas pursuing an anti-Communist policy. Moreover, the Nationalist government requested that supplies en route to North China cities be sent instead to Taiwan. The debate that took place concerning the distribution of supplies did not specifically deal with Taiwan, but by the end of December the island became one of the foci of the program.

The Economic Cooperation Administration in China, whose job it was to oversee the distribution of American economic aid there, was faced with the problem of giving commodities to areas no longer under the Nationalist government's control as North China fell to the Communists. In a telegram dated November 26, 1948, ECA mission chief Robert Lapham outlined the dilemma and spelled out the choices for action. The ECA could complete the commodity program, including publicity as to the source of supplies; continue the program by permitting the distribution of only those commodities already landed or en route to China; cancel or divert future shipments but distribute goods already landed; or cancel future shipments and attempt to reclaim goods already in China.[63]

These alternatives were debated at several Washington meetings during the following week. ECA head Paul G. Hoffman, other ECA staff members, Acting Secretary of State Lovett, W. Walton Butterworth, head of the Office of Far Eastern Affairs, other State Department members, and members of the Joint Committee of Foreign Economic Cooperation (a congressional "watchdog" committee set up by the Foreign Assistance Act of 1948) all agreed that, for the time being, the suggestion to continue the program regardless of which groups had de facto control of an area was most appropriate. Lapham's reasoning, reiterated by Hoffman, was that if aid continued to flow, Communist propaganda concerning the imperialist aims of American aid would be seen as lies; American humanitarian efforts would be demonstrated to the Chinese people.[64]

A policy statement evolved from this debate during December and received the president's approval on December 30. It stated that the implementation of the China Aid Act would continue to benefit the Chinese Nationalist government or a legal successor that pursued an anti-Communist policy. When the Chinese Communists gained control of an area, all ECA supplies that had not yet reached port were to be diverted elsewhere.[65]

What this policy meant in practice was that many of the supplies originally allocated to the Chinese mainland were diverted

Table 1

Shipments to China of Arms and Ammunition under $37.5 Million Transfer from $125 Million Grant

Ship	Sailing Date	Cargo	Arrival Date
2 LSTs	November 4 & 7 Japan and Guam	LT 1300 (approx.) small arms	Tsingtao [Qingdao] December 7–14
USS *Algol*	November 9 West Coast	LT 4974 or MT 3742 Ammunition	Shanghai Tsingtao [Qingdao] December 7
USS *Washburn*	December 1	LT 3712 Ammunition, small arms (120,000 rifles)	Due to arrive Taiwan about December 21
SS *Virginia*	November 8 Hawaii	MT 1360 Explosives	Due to arrive Taiwan about December 10
USS *Yancey*	Will Sail December 16	LT 2808 or MT 7500 Rifles, blankets, medical supplies	Due to arrive Taiwan about December 30

Source: Taken from Butterworth to Lovett, 16 December 1948, *FRUS*, 1948, 8:234.

to Taiwan. On December 7, the first shipments of military supplies purchased from the $125 million grant of the China Aid Act arrived in China. Also during that week, a request was received from the Chinese Ministry of National Defense to divert 60 percent of the shipments to Taiwan. On December 10, Acting Secretary of State Lovett notified the consul general at Taibei, Kenneth C. Krentz, that his request was approved and that arms and ammunition would be received by Nationalist personnel in Jilong, a northern Taiwan port.[66] Table 1 reveals the amounts of military supplies sent to China under the first allocations of the China Aid Act. Half of the ships were diverted to Taiwan in December 1948.

A policy of protecting Taiwan from the Communists was by no means formulated by December 1948, although it had been decided that a Communist-controlled Taiwan under the direction of the Soviet Union posed a threat to American interests in the Pacific and that China Aid Act funds would be used there. Moreover, it was not yet apparent what form of government would emerge to control Taiwan after a Communist victory on the mainland. The legal status of the island was unclear because it had been occupied by Japan during World War II. The 1943 Cairo Declaration had assured the island's return to the mainland, but at the Yalta Conference during the following year it was decided that Taiwan would not be officially returned to China until the signing of a Japanese peace treaty. This did not occur until 1951. Because of this technicality, the State Department considered alternatives to acknowledging Taiwan as a province of China, such as supporting a United Nations commission to control the island or seeking out Taiwanese independence groups.[67]

In late 1948 and early 1949 the State Department neither assumed nor hoped that the Nationalist government would take over Taiwan. On December 16, for example, the National Security Council met to discuss the Nationalists' deteriorating military situation and the decision to divert supplies targeted for North China to Taiwan. It agreed to seek additional information from the embassy in China concerning the advisability of continuing military aid and the desirability of sending military equipment to Taiwan. It would then consider "whether the military equipment in preparation for shipment to China could be more usefully employed in other theaters or whether it might be more useful to have such equipment in Taiwan provided the Chinese National Government is not established there."[68]

The embassy's reply demonstrated that Ambassador Stuart was obviously unaware of the implication that the U.S. government was planning to divorce itself from the Nationalist regime, at least when dealing with Taiwan. Stuart focused his attention on the continuation of support for Chiang Kai-shek. In his reply, dated December 18, he favored a continuation of military aid, but not its being shipped to Taiwan. Rather, it should be diverted to southern Chinese ports. He noted that should Chiang be forced to flee Nanjing, he would go to Nanzhang and then to Guangzhou (Canton). From either of these southern ports, he might rally resistance. Stuart suggested that Chiang would go to Taiwan at a much later date, and sending aid to the island at this time would only serve to encourage the evacuation of the mainland by government organizations.[69]

On the other hand, it should also be noted that additional information was received from the CIA at this time. Taiwan was a focus in the agency's December edition of the "Review of the World Situation," a document available only to selected members of the administration. The CIA concluded that

> At the moment, one element in the [world] situation stands out as of immediate and practical concern to U.S. security. This is the possible status of Taiwan (Formosa) if and when the Chinese mainland is controlled by a Communist-dominated government. On the assumption that such a development sets the stage for an expansion of the Soviet strategic position in the Far East, Taiwan, from the U.S. point of view, is strategically divorced from China and becomes one of the group of off-shore islands on which the U.S. position will then automatically rest. Soviet penetration of an island thus situated would have an adverse effect on the U.S. position on the periphery of China somewhat similar to that which a Soviet penetration of Greece would have on the Anglo-American position in the Eastern Mediterranean. There are hints that some of the Nationalist officials now being increasingly isolated in Taiwan may attempt to build up a regional authority there. The leaders of such a movement could request U.S. support on grounds that would have considerable practical appeal if a shift in the strategic importance of Taiwan had tied it in with the U.S. interests in Japan, the Philippines, and Southeast Asia.[70]

On January 14, 1949, a State Department policy statement concerning Taiwan emerged and was outlined for the president by Acting Secretary of State Lovett. It dealt with several issues: the specific question mentioned earlier of locating a naval base on Taiwan at this time; suggestions as to how the United States could maintain control of the island by utilizing the United Nations;[71] and the possible use of an American military force on the island at some point in the future. Because of the importance of this document for future policy, it will be discussed in detail.

The memorandum reiterated the recommendation that the U.S. Navy was not to be based on Taiwan. It stressed that the island was too important for American strategic interests to risk alienating the local populace. This conclusion was based on the following considerations:

> 1. The Department of State concurs in the Joint Chiefs of Staff conclusion that it is in our strategic interest that Formosa be denied to Communists.
> 2. The Communist threat to Formosa does not lie in amphibious invasion from the mainland. It lies in (a.) the classic Communist technique of infiltration, agitation and mass revolt, and (b.) the classic Chinese technique of a deal at the top.
> 3. The despatch of U.S. naval vessels and Marines to Formosa is not likely to prove effective in countering these techniques. A show of American military strength in this manner is more likely to provide Formosan fuel for the Communist fire and rally public opinion behind the Chinese Communists on the mainland.[72]

The memorandum also outlined ideas as to how the U.S. government might maintain control over the island. It noted that American military supplies were being diverted to Taiwan, and that the Chinese air force and navy were establishing headquarters there. Moreover, the evacuation to the island of families of important Nationalist officials had already begun. Thus, there were two possible sources of future conflict—a threat of takeover by the Communists and a Taiwanese revolt against the Nationalists, who were establishing themselves there in increasing numbers. The State Department was aware that a widespread revolt against the Nationalist governor, Chen Yi, and subsequent massacre of Taiwanese natives had occurred as recently as February 1947. In either case, according to Lovett, the United Nations could take action on the grounds that the situation was a threat to peace. UN intervention might be requested by the Australian or the Philippine government, citing the desire to arrange a plebiscite to determine the wishes of the Taiwanese people.[73] The Taiwanese would vote on a return of the island to the mainland or "some alternative trusteeship arrangement pending their qualification for independence."[74]

The memorandum concluded that "the Department of State fully recognized that it may be necessary at some stage for the United States to take military action if Formosa is to be denied to the Communists." It recommended that "the United States should, as it is now doing, put itself in a position to intervene with force if necessary. Such intervention should be publicly based not on obvious American strategic interests but on principles which are likely to have support in the international community, mainly the principle of self-determination of the Formosan people." Meanwhile, however, for political reasons—in Taiwan and internationally—the United States should avoid crude unilateral intervention and should look to other possibilities to establish a non-Communist government on Taiwan.[75]

An Amendment to the China Aid Act

Among the other possibilities pursued by the State Department was the continuation of aid to China after the expiration of the China Aid Act in April 1949. Discussion on this issue began in January 1949 in meetings held to consider recommendations by Paul G. Hoffman, ECA administrator. At the time of the initial discussions the Communists held Manchuria and nearly all of China north of the Huai River except the cities of Beijing, Tianjin, and Qingdao. They appeared to be in a position to gain control of Beijing, Tianjin, Nanjing, and Shanghai within a short time, and this was accomplished by early May. As a result, State Department officials concluded that by the aid act's expiration date, April 3, the Nationalist government's control would not extend beyond the southeast coast and Taiwan. It was also assumed that the Nationalist government would not be able to maintain a foothold in South China without the extension of "unlimited U.S. economic and military aid involving extensive control of the Chinese Government operations by American military and administration personnel . . . including the immediate employment of U.S. armed forces to block the southern advance of the Communists." "On the other hand, the isolated position, limited area and economic viability of Taiwan offer some prospect that a non-Communist government on Taiwan might be able to withstand Communist control of that island indefinitely."[76] One reason for this conclusion was that Taiwan had been a target of ECA-sponsored projects since the program's beginning, and after funds were diverted away from the North to southern areas, even more aid was allocated there. In January, for example, Butterworth earmarked $14,125,000 to the island for reconstruction projects. He also proposed that most of the remaining China Aid Act funds, approximately $275 million, be focused on Taiwan "in an effort to create stability which would tend to thwart Communist infiltration." But, although the U.S. government should be in a position to give such assistance through an extension of the cut-off date, it should avoid, "insofar as possible, public demonstration of a special interest in Taiwan."[77]

An amendment to the China Aid Act, proposed by the ECA and supported by the State Department, extended the act's implementation beyond April 3 to June 30, 1949. This strategy was chosen to avoid the need to sponsor new legislation for fiscal year 1950. It was considered unwise to request legislation for the new fiscal year because the degree of uncertainty and deterioration in China made it impossible to formulate viable plans.[78] In the proposal to Congress, however, the cut-off date of June 30 was mentioned merely as a target date for finishing the program. It requested authorization to use the yet unexpended $275 million from April 2 until December 31. Another modification proposed was to allow the ECA administrator to spend the funds for use by the "Government of China or such other beneficiaries in China as the President may authorize" without ratification or formal agreement. The flexibility allowed by this statement implied that aid could go to other groups in China or to Taiwan if the Nationalists remained in South China and reached an agreement with the Communists.[79]

ECA mission chief Lapham criticized this proposal and offered another alternative. In a memorandum to Hoffman, he complained that the ECA's suggestions would not provide adequate funds to carry out programs already begun in China. He asserted that if more aid were not forthcoming, the "ECA will be in a serious jam unless Commies are kind enough to occupy Shanghai in the very near future"; then American aid to that area could be stopped. Lapham summarized the ECA's financial situation in China at that time:

Estimated available April 3 [1949]	$69,500,000
Less	
Necessary earmarking, JCRR [on mainland and Taiwan]	10,000,000
Necessary earmarking, Taiwan Industrial Program	20,000,000
Commodity program on June obligation basis [on mainland and Taiwan]	61,823,000
Administrative and other	3,100,000
Total	$99,323,000
Deficit	$29,823,000

Lapham noted that "the above set-up assumes that ECA and State are serious about Taiwan." But the ECA's proposed amendment made it financially impossible to carry out its program in China, especially since Lapham felt that $20 million for an industrial program and $4 million for a fertilizer program on Taiwan might prove insufficient. He was particularly worried that the State

Department would discontinue commodity programs on the mainland to save substantial funds for Taiwan. "To do so," he concluded, "would arouse extreme suspicion and criticism of American motives with respect to Taiwan." On this point, he was assured by Philip D. Sprouse, head of the Chinese Affairs Division of the State Department's Office of Far Eastern Affairs, that capital commitments to Taiwan would be made before the commodity program on the mainland had ended in order to avoid embarrassment. Lapham then concluded by suggesting a one-year extension on the ECA program in China with an additional appropriation of $225 to $250 million. Although this amount would not be sufficient to carry on programs in operation on both the mainland and Taiwan for another year, it was noted that "in all probability the present state of affairs in China will be substantially modified within the year." Thus, any aid not yet allocated could be diverted from areas coming under Communist control to other areas in the south of China or Taiwan.[80]

Meanwhile, Senator Patrick McCarran (D., Nevada), considered an anti-administration conservative Democrat, introduced an alternative proposal for aid to China, S. 1063, which would have authorized the president to allocate $1.5 billion in aid for one year, of which $500 million was earmarked for currency stabilization, $300 million for economic assistance, and $700 million for military assistance.[81] On March 10, fifty senators, including twenty-one Democrats, sent a letter to Senator Tom Connally, chair of the Senate Foreign Relations Committee, asking that committee to support McCarran's bill.[82]

On March 15, Secretary of State Dean Acheson also sent a letter to Connally, explaining that this bill asked for aid "of a magnitude and character unwarranted by present circumstances in China" and therefore should not receive the committee's support. Acheson claimed that despite aid to China totaling over $2 billion since the end of World War II, the economic and military position of the Nationalist government had deteriorated to the point where the Chinese Communists controlled almost all of

North China. Furthermore, the Nationalists did not have the military capability to maintain a foothold in South China against a determined Communist advance. "The [Nationalist] government forces have lost no battles during the past year because of lack of ammunition and equipment, while the Chinese Communists have captured the major portion of military supplies, exclusive of ammunition, furnished the Chinese government by the U.S. since V-J Day." Acheson concluded that an attempt to underwrite the Chinese economy and military effort would be a burden on the American economy while remaining ineffective in China.[83]

But McCarran's proposal did receive support from many Nationalist sympathizers in Congress, particularly after Chinese Ambassador Koo started a campaign to refute Acheson's conclusions. The Chinese Foreign Ministry had instructed Koo to furnish such data secretly to people and institutions in the United States who were friendly to the Nationalists, including Senator William F. Knowland (R., California), Representative Walter Judd (R., Minnesota), Senator Styles Bridges (R., New Hampshire), and Senator McCarran.[84] Claire Chennault, of World War II Flying Tigers fame, testified before Congress in support of the bill.[85] Despite these efforts by Nationalist supporters, McCarran's bill was defeated, and an amendment to the 1948 China Aid Act similar to that proposed by the State Department was passed on April 19. The balance of the funds of the earlier legislation were made available to the president until February 15, 1950, "for assistance in areas in China which he may deem to be not under Communist domination."[86]

Conclusion

As a result of the passage of this bill, ECA funds remained available after the April 1949 expiration date of the China Aid Act. Although the original purpose of the act was to prop up the Nationalist government on the mainland, the Truman administration began to use the money to back its still tentative plans to defend Taiwan. Both Truman and Marshall believed that the Nationalist cause on the mainland was doomed, but the strategic importance of Taiwan, coupled with Nationalist requests to send supplies there, justified the continuation of aid. While Nationalist supporters in the United States pushed for American help in the fight against the Chinese Communists, the Nationalists began stockpiling materials on Taiwan, and the Truman administration began to look more closely at what could be done to keep Taiwan out of both Nationalist and Communist hands.

3. The Merchant Mission

The stage had seen set by April 1949 to continue aid to China through the China Aid Act with a focus on Taiwan. But, beginning as early as January, trouble was brewing for this policy. The January 14 State Department policy statement had mentioned the potential problem of local resistance to large-scale settlement by mainland Chinese on the island, and starting in January, the American consulates in Taibei, Shanghai, and Hong Kong were reporting that the local population was bitter about the use of the island as a point of retreat by the Nationalists. Information sent to Washington portrayed the abuse of the Taiwanese people, a worsening economic situation, harsh government policies, and a restive, resentful population. It was obvious to the State Department that the Taiwanese considered that American aid was facilitating the oppression of the local populace by the Nationalist "conquerors," and that another uprising similar to the one in 1947 was in the making.

This situation posed an obvious dilemma for American policy makers, which the CIA succinctly spelled out in its January 1949 edition of "Review of the World Situation." The report acknowledged that the Nationalists had already provided for adequate military control of the island and were well-advanced in their preparations for using Taiwan as a safe haven "for the last remains of Chiang Kai-shek's authority." But there were several problems that would result: Chiang Kai-shek's regime would be cut off from the resources of the mainland and would find its claims to international recognition viewed with growing doubt. "It would, however, have something to bargain with—the undeniable importance of Taiwan to U.S. strategic requirements. The fact that the National Government is asserting a legitimate authority over the island has fundamentally altered the situation in Taiwan."[1] Also, on January 28 the State Department received a communication from John M. Cabot, consul general at Shanghai, relating information he had learned in a conference with K. C. Wu, mayor of Shanghai. Wu had told him that Taiwan was planned as a last Nationalist bastion. The Nationalist military, he said, intended to withdraw to the island to defend it from the Communists, and they had already shipped 2,000,000 ounces of gold there. Wu was asking to transfer another 500,000 ounces of the 800,000 ounces remaining in Shanghai.[2]

NSC Program

A top-secret report of the National Security Council to the president dated February 3 outlined additional suggestions, including development of and support for a local non-Communist regime on Taiwan that would "provide at least a modicum of decent government for the islands." Moreover, the U.S. government "should also use [its] influence wherever possible to discourage the further influx of mainland Chinese." And it "should also seek discreetly to maintain contact with potential native Formosan leaders with a view at some future date to being able to make use of a Formosan autonomous movement should it appear to be in the U.S. national interest to do so."[3]

The report continued with an explanation of how this was to be accomplished:

> This Government should make it discreetly plain to the governing authority on Formosa that:

(a) The U.S. had no desire to see chaos on the mainland spread to Formosa and the Pescadores;

(b) The U.S. has not been impressed by the Chinese administration on the island and believes that if there is continued misrule, the Chinese authorities would inevitably forfeit the support of world opinion which might be expected to swing in favor of Formosan autonomy;

(c) U.S. support for the governing authorities of Formosa will inevitably depend in a large measure upon the efficiency of their regime and the extent to which they are able to contribute toward the welfare and economic needs of the Formosan people and permit and encourage active Formosan participation in positions of responsibility in Government.

(d) The U.S. cannot remain unconcerned over possible developments arising from the influx of large numbers of refugees from the mainland and the consequent effects, including the increasing burden on the island's economy, and is disturbed at the indication of the Chinese belief that the building up of military strength on Formosa will in itself provide an effective barrier to Communist penetration;

The U.S. government, through the most flexible mechanisms possible, should conduct a vigorous program of economic support for the economy of Formosa, designed to assist the Formosans in developing and maintaining a viable, self-supporting economy.[4]

The President approved this policy on February 4. He furthermore directed that it be implemented by appropriate executive departments and agencies which would be coordinated by the secretary of state.[5]

One aspect of the implentation of the program was the sending of a special representative, Livingston T. Merchant, to Taiwan. On February 14, Merchant, at that time counselor of the embassy in Nanjing, received a communication from the State Department ordering him to "assume special responsibility with respect to Taiwan." He was to travel to the island but was to indicate that his purpose was merely to oversee expanded American representation there, the result of Communist moves into South China. He was to avoid any implication that he had been transferred to

Taibei.[6] Dean Acheson outlined the plans for Merchant's mission to the National Security Council on February 18. Merchant was to approach General Chen Cheng, governor of Taiwan, and inform him that the U.S. government was prepared to give economic support for the island's economy to assist in developing a viable, self-supporting economy. Also, to aid Merchant's undertaking, Acheson proposed that the ECA make a study and submit recommendations for an overall program of economic assistance to Taiwan.[7]

Merchant has also described the purposes of his mission:

> My original instructions, which sent me over there and were top secret, were to establish a personal relationship with the Taiwanese underground, the belief in Washington (with reasonable justification) being that the Nationalists would not be able to get over there in sufficient force and power to dominate the island. The island, being strategically of great importance, would be run by the native Taiwanese leaders unless and until the Communists were able to mount an invasion. And my original instructions were to get over there and meet the local leaders and establish a relationship of mutual confidence with them.[8]

While Merchant was in Taiwan, other alternatives were considered in case his mission failed. One such plan was the use of outright military force to keep a Communist government from gaining hold of the islands. Acheson sought the advice of the Joint Chiefs of Staff on this issue, and on February 10 they sent a memorandum explaining why it would be unwise for the U.S. government to commit itself militarily to the defense of Taiwan at that time. Admiral Louis Denfeld, writing for the Joint Chiefs, explained that American military strength was spread across the globe, and making a relatively major military effort in Taiwan might leave the United States unable to meet emergencies elsewhere. He concluded that, for the time being, emphasis should be placed on securing Taiwan through political and economic efforts.[9] On March 1, however, this opinion was revised, resulting in a recommendation from the Joint Chiefs that a relatively small number of vessels should be stationed at island ports. Although the National Security Council rejected this suggestion, it pledged to reexamine this course of action if developments on the island justified a change in policy.[10]

In a statement to the Council on March 3, Secretary of State Acheson explained why he did not support a military show of force at that time. The U.S. government was indeed attempting a separatist movement in Taiwan, but it did not want to run up against the potential threat of an irredentist movement that would spread throughout China, especially when American propaganda was actively exploiting the Soviet-irredentist issue in Manchuria and Xinjiang. Acheson concluded that "it is a cardinal point in our thinking that if our present policy is to have any hope of success in Formosa, we must carefully conceal our wish to separate the island from mainland control."[11] On the other hand, this did not imply that the option of military force should be discarded; rather, it should be postponed until diplomatic and economic measures failed to guarantee that the island would remain free from communism. Acheson hoped that the military establishment would engage in planning in the event that it be necessary to call for a force in Formosa.[12] According to Admiral Sidney Souers, executive secretary of the National Security Council, Acheson commented that "even if we have to take it after a war begins we can't take it now." Souers also noted that the president approved of this policy.[13]

Thus, early in 1949, a policy of attempting to secure Taiwan against a Communist takeover had been worked out, although it was still not clear which formula would be successful. ECA-sponsored reconstruction projects and industrial development programs to bolster Taiwan's economy had started, military supplies paid for by the China Aid Act funds were being shipped to the island, special representative Livingston Merchant was trying to secure the confidence of both Nationalist officials there and local Taiwanese leaders, the Joint Chiefs had agreed to plan for military action should the need arise, and the use of UN intervention if a war broke out on the island was being tentatively considered. But uncertainties remained. It was unclear which group—a Taiwanese leadership or the remains of the Nationalist government—would emerge as the more powerful force. With the benefit of hindsight, it is possible to agree with the January CIA report that the Nationalists would eventually secure the island, but this was not yet apparent to the State Department or the American embassy in China.

Li Zongren Replaces Chiang Kai-shek

The role of Li Zongren was an important factor in calculating the future of the mainland. Acting president of the Chinese Nationalist government since Chiang's resignation on January 21, Li came to power with an apparent opportunity to bring about changes in the pattern of corruption, inefficiency, and backwardness of the regime headed by Chiang and his family. Li was viewed as a capable alternative to Chiang. In a memorandum written about two months after Li took over, Ambassador Stuart described the "remarkable political job" done by the acting president. According to Stuart, Li had inherited a bankrupt administration, a defeated army, and a corrupt government. He succeeded in putting Communists on the defensive with his propaganda and starting peace negotiations; he ousted several corrupt high-level Chiang appointees; he showed support for the Nationalist army still intact; and he did all this while avoiding a break with Chiang. Stuart considered this quite an achievement of leadership, which perhaps would prove formidable to both the Communists and Chiang's group in China. Li was also thought of as a friend of the West. Stuart concluded that "his usefulness to [the] cause of world peace as [an] effective instrument for [the] containment of communism in [the] Far East should not be underestimated."[14]

The importance of Taiwan to the United States was again demonstrated during Li's brief tenure as acting president. His secret talks with the Communists were surrounded by rumors concerning a deal involving the island if a coalition government could be formed. The State Department did not want Taiwan to fall into the hands of a coalition government of which the Communists would be a part, and as a result, in late March Merchant was informed that he must familiarize himself with all independence groups even if they were weak so that they might be used to stimulate UN action to protect the island. To this end, Philippine authorities had already been informed of America's "general interest" in the island and their role in initiating any UN action necessary. There remained "no thought that the United States government would act unilaterally" to separate the island from the mainland by military means.[15]

Merchant Favors the Nationalists

These plans to support independence groups never reached fruition because the American outlook toward the Nationalists on Taiwan had changed by April 1949. Merchant apparently spent most of March searching for effective leaders with whom to work. Local Taiwanese leadership was weak because several resistance movements were located in Hong Kong and Shanghai and not on the island. Furthermore, Governor Chen Cheng, a Chiang appointee whom Merchant had initially been told to seek out, was lackluster and ineffective. In March Merchant suggested to the State Department that the U.S. government use its influence to persuade Li Zongren to replace Chen with a more powerful Nationalist official, for example Sun Liren, commander in chief of the Taiwan Defense Headquarters and deputy director of military and political affairs for Southeast China.[16] The debate on this issue continued, and Chen learned of the plan to remove him. He then traveled to Nanjing to convince Li and Stuart that he was doing a satisfactory job. By the beginning of April, Merchant conceded that Chen was the authority with whom the United States would work. On April 6, Merchant began to negotiate with Governor Chen for additional American aid to Taiwan.[17]

Moreover, it was at this time that Merchant suggested a change in policy to the State Department. In a letter dated April 12, he explained that the presence of the Nationalists, particularly Chen and the governing group, served American interests in Taiwan for the time being. He had concluded that Chen and Chiang would refuse any form of Communist-dominated government on the island. Acheson agreed with this theory and told Merchant to approach T. V. Soong [Song Ziwen], an important Nationalist official and brother-in-law of Chiang, to tell him that the United States would give "primary consideration" to economic and self-strengthening projects carried out by the Nationalists. Merchant was told to be careful, however, not to inform Soong that the United States had designs on the island.[18]

These instructions were timed to coincide with Soong's week-long visit to Taiwan beginning April 11. Soong had orders from Chiang to discuss with the United States consul at Taibei, Donald D. Edgar, a political, economic, and military formula to prevent a Communist takeover of the island. He had informed Edgar that the Nationalists were planning to save the island for themselves, but they needed American aid. Negotiations were begun on an aid program, and by the end of April, $500,000 worth of wheat and cotton was sent to Taiwan, a $17-million project by the Joint Commission on Rural Reconstruction was begun, and an American engineering group was stationed there to work on reconstruction projects.[19]

It can be concluded that by June 1949, when Merchant's mission was completed, the United States government was resigned to the fact that Nationalist leaders would set up a government on Taiwan. On May 6, Edgar had informed the State Department that the Sun Moon Lake quarters in Taibei were being prepared for Chiang and his entourage. Troops and official refugees continued to pour in. Over 50,000 troops from Shanghai's 52d, 54th, and 99th armies arrived in Taiwan in early June. The local military was forced to take a backseat to the new arrivals. By June 23, it was also clear that the Nationalists were aware of the defensive value of the area to the United States and believed that American strategic requirements would compel the U.S. government to keep Taiwan in the hands of a non-Communist administration.[20]

Merchant described the influx of the Nationalists during his five months' stay on Taiwan:

> I was frankly amazed at the number of troops and the amount of military equipment and material they were able to move from the mainland, from Chinchow [Jinzhou] and from Shanghai, and get on the islands. The island was loaded with the best units of the Chinese Nationalist army, and ample material—artillery, munitions, aircraft, and so forth. And the morale with the Chinese Nationalists was surprisingly good. They felt that at least they had put a 90 mile moat of water between them and the Communists. And the Taiwanese, who had been up to that point an uncertain element, had no thought for trying to buck the Nationalists; they were docile, easily controlled, managed, and so forth.
>
> So, when I left [in June 1949], it looked really as though the Nationalists were organizing, pretty effectively, Taiwan as a fortress. And with the Communists having had no experience in amphibious landings, it looked as though Taiwan would be an awful tough nut for the Communists to take. So, when I left, I was reasonably satisfied that the Nationalists were going to be able to hold out for a very considerable period of time on Formosa.[21]

Moreover, by the time Merchant left, Chiang Kai-shek himself had arrived on the island.

Problems among the Nationalists

The situation on Taiwan was insecure for several reasons. The reactionary nature, corruption, incompetence, and unpopularity of the Chiang regime on the mainland were being transferred to Taiwan. This was one of the reasons why the State Department had sought earlier to keep the Nationalists off of the island. As troops and officials continued to arrive, peace was endangered because of the restlessness of the locals. Reports were received that the Nationalist officials and civil servants were "yes men" of the unpopular Governor Chen, and his government was described as military and reactionary.[22]

Another problem for the State Department was the precarious situation within the Nationalist government itself. The Chiang-Li relationship had developed into a feud. This supported the idea that the Nationalists would fail to maintain a foothold on the mainland because Chiang still controlled much of the military as well as the government's revenue. On April 19 Butterworth presented a report to Acheson at the latter's request, outlining the obstacles confronting the Nationalist government under Acting President Li. Citing a series of messages received from American consulates in China, Butterworth noted that Li had failed to improve the military situation in China because Chiang and several provincial leaders would not relinquish their authority to him; Chiang had scuttled a Nationalist plan for defense of the Yangzi River area by deploying most of the Chinese navy and air force to Taiwan; and Li had been unsuccessful at obtaining control of the government's gold and silver reserves, held by Chiang.[23] By May it was known that the Nationalists would remain split into two groups. At this time the mainland Nationalist government, headed by Acting President Li, was situated in Guangzhou (Canton), but Li planned to move west with the remainder of the government in the face of Communist advances. Chiang would not change his plans to make Taiwan his last bastion, with part of the government accompanying him there. When Li presented his scenario to Lewis Clark, minister-counselor of the embassy in China, he inquired about the possibility of American recognition of Chiang's group on Taiwan and the abandonment of his own supporters on the mainland. Clark replied that at this point the United States would not be committed either way.[24]

The "bickering and jockeying for position" ended with Chiang's move to Taiwan and the assertation that Li would resist the Communists until all hope was gone since his effort to reach an accommodation with them during the peace talks had failed.[25] This statement apparently satisfied Chiang and his more conservative associates, who opposed any link with the Communists and had been upset by Li's attempts to negotiate.[26] But, although Li's position became somewhat more secure within the Nationalist party, his situation seemed doomed to Ambassador Stuart in China. After one of Li's many requests for American military aid and a U.S. statement of support, Stuart wrote in May to Acheson that "With all my sympathy for Li Tsung-jen [Zongren] and his determination to resist Communists, I cannot envisage any statement [the] U.S. government could make at this late date which would be effective in changing [the] course of military events, which could avoid embroiling us further in Chinese political confusion."[27]

During the summer of 1949, Li began to organize local resistance to Communist advances toward Southwest China. He made plans to give increased authority to local governments in the hope of forming a will to resist. The difficulty, according to Li, was a lack of funds—local governments could not afford to pay the cost of an army. Li considered two possible sources for such aid. First, while visiting Chiang in Taiwan in July, he requested more arms and ammunition for the troops still on the mainland. Chiang replied that all supplies had already been allocated and none would be released to Li. He also asked for money from the Nationalist treasury, but he was told that funds had to be conserved so as to last two years. Li admitted that he had been stymied in his efforts to get something for the effort on the mainland from Chiang.[28]

A second source of aid was the United States. Li reported his problems with Chiang and inquired whether the U.S. government would assist autonomous areas in South China against Communist expansion. In June one of Li's representatives, Gan Jiehou, arrived in the United States on a mission to obtain financial assistance, military aid, moral support, and advice for the Nationalists on the mainland. Gan outlined a plan for the defense of South China. With the failure of the negotiations, it appeared inevitable that the Communists would continue their march across the Yangzi River and take all of South China. In late April the Communist armies had begun to cross the Yangzi River almost unopposed. To hold Guangdong in the Southeast and adjacent provinces to the west, Gan said, arms, silver, and troops were urgently needed. He reported that Li was prepared to denounce Chiang and oust his associates from South China, make any necessary reforms, and provide an effective defense if American help in these matters were forthcoming.[29]

The State Department replied to these proposals by reminding Gan and Li that South China was already receiving "substantial material support from the United States through the ECA," especially since millions of dollars orginally earmarked for North China had been diverted to those areas not yet under Communist control. Moreover, it was impossible for the U.S. government to underwrite the kinds of changes Li's government proposed because of the implication of a long-range commitment.[30]

Conclusion

The State Department's negative responses to inquiries for increased aid by the Nationalists on the mainland in the summer of 1949 and its acquiesence in, if not support for, allowing substantial amounts of economic and military aid to be targeted to Taiwan reveal that the conclusion had already been reached that nothing could be done to prevent the Communists from defeating the Nationalists on the mainland. Taiwan, on the other hand, was considered seriously as the site for an anti-Communist stronghold. This meant that the State Department would be faced with the existence of two claimants to the government of China—the Chinese Communist Party, which already held de facto control over most of the mainland provinces, and the representatives of the Nationalist government, who had fled to Taiwan.

By sending supplies to Taiwan, the United States was aiding the Nationalists not in a fight against Communists, but in setting up a rival government. As a result, the Merchant mission was a failure since it did not offer the Truman administration a way to hold on to the strategically located island without supporting an unpopular regime.

4. Problems for Americans on the Chinese Mainland

While Chiang Kai-shek and much of the Nationalist government were consolidating their position on Taiwan during the summer of 1949, the world waited for the final victories on the Chinese mainland that would augur a new Chinese government. By May the State Department was convinced that most U.S. allies, including the British, planned to grant the Chinese Communists at least de facto recognition relatively soon. The British wanted to protect their substantial economic interests on the mainland, most notably the colony of Hong Kong. The State Department also had been considering the question of the recognition of a Communist government for about a year. This volatile issue stimulated a variety of problems because of vocal support for the Chinese Nationalists and a "tradition" of helping Chiang since World War II, on the one hand, and American economic interests and competition with the British for trade in China, on the other. But the most pressing conflict, which Secretary of State Dean Acheson has referred to as the one that prevented recognition, was the November 1948 arrest of the American consulate staff in Mukden by Chinese Communist authorities and their detention until December 1949. For over a year, this situation colored American relations with the CCP, the British, and other allies.

Americans Held Hostage in Mukden

On November 2, 1948, Chinese Communist forces occupied Mukden (Shenyang[1]), a city in the nothern province of Manchuria. The first few meetings between Americans there and the new authorities went smoothly, but conditions deteriorated rapidly and Chinese soldiers were placed on guard at consulate headquarters, presumably for the Americans' protection but, according to a Chinese reporter, actually to observe American activities. On November 18 the consulate's radio transmitter was seized and, after Consul General Augus T. Ward tried to refuse to surrender it, all "U.S. citizens in Mukden were interned in their houses." From November 20 the consulate staff was under house arrest.[2]

The State Department was naturally disturbed. Ambassador John L. Stuart in China suggested that a "strong blast from an official spokesman" of the department would help teach the Communists "more correct international manners.[3] The State Department felt, however, that a unilateral statement would be ineffective. What was needed was a joint approach by all governments with consulates in Mukden. Therefore, Stuart was authorized to question the British and French embassies in Nanjing to get their reaction to this proposal. By January 12 he had drafted a joint statement on the Communists' denial of communication facilities in Mukden, but he was unable to sell the idea to either the British or the French. Both thought that such a statement would hurt more than help their cause of preserving their economic and political interests in China. As a result, Lovett suggested that the statement not be issued.[4]

The motives behind British and French concern for their interests in China involved their attempts to deal with the question of recognition of the new Chinese government once the Communists came to power. It was clearly only a matter of time before the Communists would gain control of most of the mainland and announce the establishment of a government; local Communist administrations had already been set up in North China. As a result, discussions took place among representatives of the Western allies in order to develop a concerted action toward the question of recognition. During a meeting on this subject between Philip D. Sprouse, chief of the Division of Chinese Affairs, and J. F. Ford, first secretary of the British Embassy, Sprouse told Ford that the question of recognition merited careful study because of its possible use as a weapon to protect American interests and property in China. The State Department, he said, was approaching only the British for the time being, but it would also be desirable to have a common policy with the French, since all three had interests in China "which could be served advantageously in this connection."[5]

American policy on this issue had already been formulated several months earlier. In July 1948, at the request of Secretary of State George C. Marshall, the office of Intelligence Research had prepared a seventy-nine-page report entitled "Problems of Domestic and Foreign Policy Confronting the Chinese Communists," which outlined American objectives in East Asia for the following five years. These objectives fell into four categories: (1) maintaining and developing conditions in East Asia favorable to a non-Communist solution of East Asian problems; (2) encouraging differences between the Chinese Communists and the USSR; (3) maintaining an island security belt off the coast of East Asia; and (4) promoting freedom of international trade.[6] It was through freedom of trade that the State Department saw a possible route toward developing relations with the Chinese Communists. The report concluded that "despite China's strong political attachment to the Soviet orbit, the need to realize the maximum possible return from foreign trade may lead Communist leaders to seek close economic relations with non-Communist areas." It would be a disaster for the Chinese economy, for example, if relations with non-Communist countries were minimal. On the other hand, a unilateral U.S. refusal to trade would have less effect.[7] During the following months, the State Department consistently upheld a policy of maintaining trade and other economic ties. This served a dual purpose of offering a choice of trading partners for the Communists as well as protecting American economic interests in China.

Meanwhile, the State Department outlined the manner in which the embassy in China should deal with the question of recognition. Because of continued recognition of the Nationalist government and the Communist practice of addressing American embassy personnel as private citizens, Acting Secretary of State Lovett instructed the embassy to "exercise every caution [to] avoid any step which might be considered as leading or tantamount to recognition." The embassy was to refrain from visaing passports issued by the new regime and addressing any formal communication to its leaders in their official capacity that might be construed as constituting recognition.[8]

The State Department maintained close contact with the British Embassy in Washington, and in January Sprouse met again with Ford several times to discuss the question of recognition.[9] On January 5 the British ambassador, Oliver S. Franks, sent Lovett a copy of a memorandum drawn up by Ernest Bevin, British foreign minister, for consideration by the British cabinet. Franks requested that Lovett nominate a member of the State Department to confer with the counselor of the British embassy, Hubert A. Graves, on the details of the memorandum. The document outlined a plan to deal with the rapidly developing situation in China and came to the following conclusions:

> In China it can be assumed:
> (a) that there will be an immediate period of dislocation when foreign commerce generally will be at a low ebb;
> (b) that there will follow a period in which the economic difficulties of the Communists may dispose them to be tolerant towards foreign trading interests;
> (c) that the present nationalist tendency towards foreign investments and capital installations will thereafter be enhanced and that the intention to work rapidly towards the exclusion of the foreigner will be strengthened;
> (d) that there would be a tendency to subject foreign trade, both import and export, to close government control, which

would not altogether suit the types of trade United Kingdom merchants aim at doing in and with China.

It recommended

(a) that His Majesty's Government should consult with the Government of the United States, the British Commonwealth, France, Netherlands, Burma and Siam as to the best means of containing the Communist threat to our several interests.
(b) that all necessary steps should be taken to strengthen our position in colonial territories in the area.
(c) that we should consider, in consultation with friendly Powers whether the economic weakness of Communist-dominated China might not offer an opportunity to secure reasonable treatment for our interests.[10]

On the following day, Sprouse and Butterworth met with Graves to clarify the British position on the question of recognition as outlined in the memorandum. Graves assured the Americans that the British would not automatically recognize a successor government without first carefully studying its character, taking into consideration "the extent of its control and the manner in which it would deal with British interests and trade."[11] He reiterated, as had been done at the other January meetings, that Americans and British had similar views on the issue of recognition. But the State Department continued to be troubled by the situation in Mukden.

Consideration was given to threatening withdrawal of American personnel from Mukden and Tianjin, where Communist authorities on January 23, 1949, had confiscated another consular radio apparatus, one week after they occupied the city. But both O. Edmund Clubb, consul general in Beijing, and Lewis Clark, minister-counselor of embassy in Guangzhou, felt that this might play into Communist hands because it was probably what their Soviet advisers desired. Clubb and Clark felt that by threatening to leave the consulates, the United States might find itself with "listening posts" only in the capital, as was the case in the Soviet Union. Thought was also given to sending messengers, either Chinese or American, to observe conditions in Mukden. Throughout February State Department and consulate staff members in China continuously expressed apprehension over the well-being of the Mukden staff. On March 2 Acheson told Clubb to deliver a strong statement to the "highest Commu[nist] authority available" indicating that the United States government was seriously concerned for the welfare of its people and that the "arbitrary restrictions" placed on the consular officers in Mukden were in total disregard for international custom.[12]

About this same time, the State Department received word that the British planned an independent policy of recognition. A memorandum handed to Sprouse by Ford on March 21 outlined the views of the Foreign Office. With most of North China to the Yangzi River under Communist control and prospects for the formation of a coalition between the Communists and Li Zongren dim, the British had concluded that to refuse to recognize a government that controlled a large proportion of territory was not only "objectionable on legal grounds but leads to grave practical difficulties." It was a possibility that the British government would recognize the Chinese Communist government as the de facto government of the part of China under its control while continuing to recognize the Nationalist government as the de jure government of the whole of China. The memorandum indicated, however, that the British would proceed only after consultation with other powers.[13]

The State Department continued to attempt to resolve the Mukden conflict. On April 15 Acheson requested Clubb in Beijing and George D. Hopper, consul general in Hong Kong, to send a communication to the Communist authorities. They were to refer to the fact that no reply had been received from the earlier message sent in March and repeat that the blocking of communication with Mukden had created an intolerable situation. The reported confinement of the consular staff was "so clearly contrary [to] universally accepted principles [of] international comity" that the U.S. government could no longer delay clarification of the situation. During that month more disturbing information had been received from the consulates in China. There had been no direct word from Mukden for over five months, and blind broadcasting from Shanghai to Mukden received no confirmation. The British consul in Mukden sent a letter to Beijing saying that "foreigners (except Americans) have . . . no personal ill treatment to complain of." Also, a "white Russian" who had recently returned from Mukden informed the consulate at Dalian that guards surrounded the American consulate and residences but they were also at the Soviet consulate. He also said that the American flag was still flying over American buildings. Hopper was unsuccessful in contacting the appropriate Communist authorities in Hong Kong because of their refusal to permit official contact with representatives of the U.S. government, which did not recognize the Communist authority. He tried to relay the letter through a Chinese Communist representative in Hong Kong but was unsuccessful. Attempts to get permission to allow two Americans to travel to Mukden also failed.[14]

These problems served as background to the frequent reports from the British that they would soon recognize the Communist authorities as the de facto government in areas under Communist control to prevent disruption of British affairs in China. By April, British prestige had decreased as a result of a series of incidents involving British traders and Chinese Communists in the Yangzi River area. On April 30 a Foreign Office spokesman in Shanghai announced to the press that the British government planned to establish "friendly relations" with the Communist government. This move coincided with a directive from Acheson informing Communist authorities in Mukden that unless the situation there was rectified, the United States would be forced to withdraw its consular personnel. On April 30 a registered letter containing that ultimatum was sent to the General Headquarters of the People's Liberation Army in Beijing.[15]

The British intention to recognize the Communists was not well received in the State Department or the embassy in China.

Cabot was one of the few who saw the merits of this idea regarding the protection of interests in China. Stuart and Lewis Clark, minister-counselor of embassy at Guangzhou, both believed that the Western countries should wait until a formal government had been set up and requested recognition. Stuart felt that the Communists were succeeding in "playing off one foreign power against another" and that the North Atlantic community, particularly the British, "should not be permitted to jump the gun; for temporary apparent commercial political gains which [the] CCP may well attempt to dangle."[16]

Stuart sent telegrams noting his views to members of the North Atlantic community and also expressed his opinion personally to the British ambassador in China, Ralph Stevenson, on May 4 during his "first outing" in a week because of Communist control over his activities since their takeover of Nanjing. On the following day, Stuart met again with Stevenson and French ambassador Jacques Meyrier, both of whom stressed the practicability of granting de facto recognition given the rapid disintegration of the Nationalist government and the need to arrange for satisfactory protection of their countries' interests in China. Stuart countered by saying that they should not sacrifice long-range advantages for immediate and relatively minor ones.[17]

On May 6 Acheson, in support of Stuart's views, sent a telegram to American consulates in London, Paris, Rome, Brussels, the Hague, Ottawa, Lisbon, and Canberra authorizing consular officials to discuss the question of recognition with the foreign minister in their host country. They were to emphasize the disadvantages of initiating any moves toward recognition or giving an impression that any approach by the Communists would be welcomed, and they were to stress the desirability of adopting a common front on this problem among the Western powers.[18]

Within a week Acheson also prepared a policy statement concerning the issue of recognition of the Communists in China. The policy was cabled to Stuart on May 13. The State Department's position was that the recognition of any new government should be based on the following factors:

(a) *de facto* control of territory and administrative machinery of State, including maintenance [of] public order;

(b) ability and willingness of gov[ernmen]t to discharge its internat[iona]l obligations;

(c) general acquiescence of people of country with gov[ernmen]t in power.

Furthermore, recognition by U.S. sh[ou]ld not be withheld as [a] political weapon except in extreme cases when U.S. nat[iona]l interest served thereby.

Re: [the] question [of] Commies as in *de facto* as opposed [to] *de jure* control, [the] foll[owing] considerations appear pertinent:

(a) US Gov[ernment] on recent occasion (Israel) recognized provisional gov[ernmen]t as *de facto* authority and exchanged representatives prior to *de jure recognition.*

(b) *Recognition of de facto* authority can legally be extended without withdrawing recognition from *de jure* gov[ernmen]t. (Oppenheim's *Internat[iona]l Law*, Vol. I., 7th Edition, pp. 145–146 par 75g)

(c) Granting recognition *de facto* authority Commie regime would politically encourage Commies and discourage Nat[iona]l Gov[ernmen]t.

(d) When Phibun Govt. succeeded Khuang Govt. in Siam, we asked and obtained assurances of intentions [to] fulfill internat[iona]l obligations and accordingly did not withdraw recognition. US has withheld recognition from Albania in view [of] its refusal [to] give similar assurances.

The cable concluded:

the Commies have not as yet established [a] "central gov[ernmen]t" in any sense of [the] word and are not seeking recognition. The Nat[iona]l Gov[ernmen]t [is the] only Gov[ernmen]t in China which has claim [to] recognition. Consular establishments [in] Commie controlled areas remain open and functioning where possible as repeatedly indicated. As functioning bodies they are on [a] practical basis handling purely local problems with local authorities. Facing facts it [is] only natural

[to] expect their relationships sh[ou]ld be as amicable as conditions permit.

Dep[artmen]t continues of opinion [that] we sh[ou]ld strongly oppose hasty recognition [of] Commies either as *de facto* or *de jure* authority by any power and sh[ou]ld continue our efforts [to] obtain full agreement [of] concerned fo[reig]n powers (particularly British) to desirability [of] presenting [a] common front [on] this question.[19]

This policy statement reflected the State Department's goals as well as the obvious dilemma of what to do about the poor treatment Americans were receiving at the hands of Communist authorities in several areas. On the one hand, the State Department wanted to assure that no moves were made by the American embassy or those of its allies that symbolized a desire to recognize Communist authority. On the other hand, the policy acknowledged that recognition of a Communist regime might be possible once a government was established or when Communist authorities fulfilled international obligations such as respecting diplomatic immunity. The department's views were further complicated by information received from Beijing which implied that the situation in Mukden should not represent the general Chinese Communist attitude toward Americans. There existed sufficient evidence for the embassy to conclude that the Soviets in Manchuria were behind the Mukden problems because they were interested in maintaining a hold on the area.[20] Clubb also reported that Zhou Enlai, a member of the CCP Central Committee, said that the Communist attitude in Mukden vis-à-vis foreigners was mistaken. He contended that the CCP should incline toward the United States rather than the Soviet Union. The anticipated arrival in Beijing of Zhou and Mao Zedong, CCP Central Committee chairman, might bring new developments for foreigners, particularly Americans.[21]

But conditions in Mukden did not improve. On the contrary, for the next several months the situation there steadily deteriorated. Communist authorities there failed to respond, and when Stuart and Clubb attempted to contact higher Communist representatives such as Mao Zedong and Zhou Enlai through intermediaries such as Huang Hua, director of the Aliens Affairs Office, and Philip Fugh, Stuart's Chinese personal secretary, they were informed that not much could be done for the Americans.[22]

But on June 1 a development occurred that supported the feeling of some Americans in China that the Mukden incident did not reflect the views of all CCP members. The embassy at Beijing received a communication allegedly from Zhou Enlai. According to a "reliable intermediary," Australian journalist Michael Keon, Zhou wanted a message transmitted to the "highest American authorities on top secret level without his name being mentioned." In his communication he outlined a brief description of several factions within the CCP and stated that there were a few temporary disagreements within the party during the present stage of the revolution, but these would be worked out during the next stage of reconstruction. He claimed that the CCP was split into two main groups, the radicals who desired an alliance with the USSR and the liberals, of which he was one, who wanted to work out good relations between China and the United States. He gave three reasons why the United States should aid China: "(1) China [was] still not Communist and if Mao's policies are correctly implemented may not be so for a long time; (2) democratic China would serve in [the] international sphere as [a] mediator between Western Powers and [the] USSR; (3) China in chaos under any regime would be [a] menace to peace [in] Asia and [the] world." In conclusion he added that he was confident that Mao Zedong would listen to the various sides and translate the ideas into practical working policies. "He hoped [that] American authorities . . . would believe [that] there were genuine liberals in [the] party who are concerned with everything connected with [the] welfare [of the] Chinese people and peace in our times; rather than doctrinaire theories."[23]

Acting Secretary of State James E. Webb related the contents of Zhou's message only to top-ranking members of the China staff. Every precaution was taken to ensure that receipt of the message was kept secret. Clubb, although not sure what might have motivated Zhou's action, said that he had foreseen the possibility of the development of Chinese "Titoism." The reason for this, he claimed, was that the Chinese economy was suffering severely, but until such time that a clear break with the Soviets occurred, it must be assumed that the CCP would remain allied to the USSR. Stuart said he wished it were true that Mao was emulating Tito, but he felt that the CCP represented doctrinaire Marxism. It seemed to him that Mao's relationship to the Soviet Union was that of a "brilliant disciple if somewhat junior ally in [a] 'world anti-imperialist front.'" Stuart also sent a second reply in which he added that he would be trying to keep in personal contact with Zhou because he felt that the State Department should take "full advantage" of the demarche, while being careful "not to over-play" any chances. Clark commented that "we should be most suspicious [of the] Chou En-lai [Zhou Enlai] demarche. Move smacks much more of tactics than strategy."[24] In a meeting with the president on June 16, Acting Secretary Webb discussed the message and the embassy's reactions. Truman directed the State Department "to be most careful not to indicate any softening toward the Communists but to insist on judging their intentions by their actions."[25] The State Department made no specific reply to Zhou's alleged communication.

Several days later, on June 11, Clubb in Beijing received the first communication from Mukden since November 20. In a brief message, Ward related that the restrictions placed on embassy personnel on November 20 remained unchanged. He had attempted to contact the Mukden mayor since November 9 to ascertain whether permission would be granted for the evacuation of his staff. Despite repeated requests, he had received no reply. He also had tried unsuccessfully to purchase packing materials.[26]

Any optimism concerning this breakthrough in communications or its possible relationship to the Zhou demarche was short-lived because within a week the English-language service of the Communists' New China News Agency published a lengthy article announcing the disclosure of a "big American spy ring" centered at the American consulate and the United States Army liaison group in Mukden. The Americans were accused of carrying out "anti-Chinese people plot activities for [the] destruction of [the] Chinese people's revolutionary enterprise and world peace." The article concluded by promising legal action against the offenders.[27]

Another significant move on the part of the Chinese Communists, however, closely followed the Zhou demarche. On June 28 Huang Hua met with Ambassador Stuart to invite him to Beijing before his return to the United States for a consultation in July. Stuart had been president of Beijing's Yanjing University from 1919 to 1946, and Huang had been one of his students. The ambassador customarily visited the university annually on his birthday and for commencement in June. Stuart regarded Huang's approach as a "veiled invitation from Mao and Zhou to talk with them while ostensibly visiting Yenching [Yanjing]." Stuart told Acheson that "to accept would undoubtedly be gratifying to them, would give [him a] chance to describe American policy; its anxieties regarding communism and world revolution; its desires for China's future; and would enable [him] to carry to Washington [the] most authoritative information regarding CCP intentions. Such a trip would be a step toward better mutual understanding and should strengthen [the] more liberal anti-Soviet element in [the] CCP. It would provide [a] unique opportunity for [an] American official to talk to top Chinese Communists in [an] informal manner which may not again present itself." On the other hand, Stuart noted that such a visit "would undoubtedly start rumors and speculations in China [of an American approach to the CCP] and might conceivably embarrass [the State] Department because of American criticism [at home]." Stuart wanted directions from Acheson concerning a response to Huang.[28]

Several State Department members reacted to this information by supporting Stuart's request for permission to go to Beijing. John P. Davies of the Policy Planning Staff wrote to the staff's director, George Kennan, reiterating a conversation he had had with Walton Butterworth of the Office of Far Eastern Affairs. Both Davies and Butterworth considered the invitation even more important than the demarche by Zhou. The utility of such a visit, according to Davies, was what Stuart could tell the Communists under State Department direction. But both Butterworth and Davies were worried about a reaction by Nationalist supporters in the United States. Butterworth suggested that, to make the trip worthwhile, Stuart should accept only on the basis that he fly in his own plane to pick up the Americans in Mukden, stopping off on the way to Beijing. Davies agreed with Butterworth "that this would make a lot of face for us in Asia and that it would be a justification in the eyes of the American public for the visit." The State Department could also make it clear to the American public that Stuart had not gone to Beijing "to play footy-footy with the Communists but had gone there . . . to read them the riot act."[29] Cabot, consul general at Shanghai, telegrammed Acheson urging that Stuart be allowed to go to Beijing to work out a system of protection of American business interests in China.[30]

On July 1 Stuart received a communication from Acheson stating that "under no circumstances" should he make a visit to Beijing. The telegram concluded that the reasons for the negative decision were those Stuart had mentioned in his telegram.[31]

Cabot has speculated about why this response was made. His "own personal interpretation" while he was in Shanghai was that the Truman administration was afraid of American public opinion, which was strongly anti-Communist. When he returned to Washington later in 1949, he received another explanation from Butterworth. Some members of the State Department did not believe that Mao and Zhou really intended to receive Stuart; their point was to get him up there to humiliate him. Later Cabot heard yet another opinion from Lewis Clark, minister-counselor at the embassy in 1949, which was that some State Department members did not trust Stuart "to say the right things at the right moment, and they didn't dare put him into a very tricky situation like that."[32]

Cabot had also predicted that a negative response to Huang's invitation would hurt American interests in China. His opinion proved correct, for U.S. relations with the Chinese Communists further deteriorated during the summer of 1949.[33]

On July 6 another incident occurred which, although not so serious or prolonged as the situation in Mukden, nevertheless represented for the State Department the nature of American relations with the Chinese Communists. Vice Consul at Shanghai William M. Olive, described by Consul General Cabot as "young, very innocent, and not too bright," was arrested, detained for three days, and beaten by Shanghai police. Olive was arrested after allegedly driving his jeep down a street that had been reserved for a parade. An Indian guard, detained at the police station during Olive's arrival, gave an eyewitness account of the beatings: When Olive appeared for questioning, he was argumentative and hit the counter with his hands several times. Then several police officers entered to pull him away. Olive reached for the counter, pulling a letter box off the counter and splashing ink on his own face and clothes as well as those of the questioning officer. Blows and kicks were exchanged between Olive and the other officers and he was handcuffed, whereupon the officers continued to beat him. One officer drew his gun and told Olive to get into a cell, but Olive refused. The officers then threw him in bodily. The police later refused to let Olive use the telephone.[34]

Attempts by members of the consulate staff to visit Olive and bring him food during the next two days were unsuccessful. It was learned that the police had been making persistent efforts to induce Olive to apologize, but he steadfastly refused to do so. On July 8 Cabot consulted a "close foreign contact" who agreed to work on the Americans' behalf. After meeting with the Chinese police, the contact was pessimistic because Olive's obstreperous attitude complicated the case to the extent that he might be detained for months before a sentence could be decided. The contact also suggested that Olive's wife write to the commissioner of police and deliver in person a sincere apology for what her husband had done. He commented that the apology should be written in a sentimental way in which she declared her love and concern for her husband. In the meantime, Ambassador Stuart in Nanjing attempted to meet with Huang Hua of the Alien Affairs Bureau but was rebuffed since the Chinese would accept no protests from any foreign governments with which they had no diplomatic relations.[35]

On the evening of July 8, a Shanghai police officer visited Mrs. Olive in her apartment, apparently in response to her letter. Olive himself, earlier that day, had signed an apology admitting "serious errors." The police officer had come to pick up some clean clothes. On the next day Olive was released in the custody of his wife after paying a fine. Upon hearing the outcome, Secretary of State Acheson suggested that Cabot lodge a protest in conjunction with other foreign consulates.[36]

While the well-being of the Americans in Mukden, Shanghai, and elsewhere in China appeared to be in jeopardy during the summer of 1949, Washington was also receiving more reports from its allies that they were willing to recognize the Communists as de facto authorities in areas under their control. The French acknowledged that they were determined to normalize relations with the Communists at the earliest date in order to protect their interests in Southeast Asia. The Australians also were firm in their attitude that they would recognize the Communists as soon as possible, and they condemned what they considered to be an attempt by the United States to erect an "economic cordon" around North China. The British prepared a lengthy memorandum on August 17 reviewing the situation in China and assessing British mercantile and missionary interests there. The Foreign Office concluded that withholding recognition from a government in control of a large part of China would lead to grave practical difficulties concerning the protection of Western interests in China.[37]

Another, related phenomenon that also presented problems for the State Department were the various reports received from Mukden explaining that while the Americans remained "complete prisoners," the British and French positions there were continually improving. Guards had been withdrawn from these consulates, telephones had been restored, and the foreigners' automobiles had been returned. This prompted the American counselor of embassy in Nanjing, John W. Jones, to comment, "Why [the] Commies are showing such relatively favorable consideration to British and French Con[sulate] Gen[erals], given Soviet-type iron curtain which has descended around Manchuria, we are unable to explain."[38] One reason for the favorable treatment involved British and French trade with the CCP. This was yet another source of conflict between the United States and its allies.

American and British Competititon in China

American policy toward trade with China had come under fire from several branches of the government and from the president at the time of Ward's arrest because American businesses had continued to trade with the Communists with the knowledge and approval of the State Department. In early November 1948, Truman ordered Acheson to begin a revision of this policy.

The U.S. trade policy toward the Chinese Communists was spelled out in February 1949 by the National Security Council. The NSC's goal was "to prevent China from becoming an adjunct of Soviet power"; to do this, it was determined that "ordinary economic relations between China on the one hand and Japan and the Western world on the other" must be maintained. Such a plan solved two problems for the United States. It would ease the burden of American expenditures in Japan and allow for the continued operations of private American interests in China, including such American-owned businesses as the Shanghai Power Company, which supplied that city with most of its electricity. The State Department had opposed suggestions by company officials that the plant be abandoned once that city came under Communist control. John Cabot, consul general at Shanghai, told the American head of the company that it might be in the United States' best interests for the power company to render its services even after the "last ship" left Shanghai. The plant could serve as a "toe-hold" in Communist-occupied China.[39] Moreover, the importance of such economic relations to the Communists "might foster serious conflicts between Kremlin and Chinese Communist policy," which could lead to an independent Chinese Communist regime. The United States could adopt a more restrictive trade policy at a later date if the Chinese were to challenge American strategic interests. The report also stated that some restrictions were to be placed on the kinds of items traded. To prevent the Chinese from re-exporting certain strategic items to the Soviet Union or its spheres of influence in Eastern Europe and northern Korea, controls would be placed on exports of military commodities and certain industrial, transportation, and communications supplies.[40] Truman approved this policy on March 3 and directed that the appropriate government agencies implement its conclusion.[41]

Problems arose for the new trade policy in early June when the Nationalist government announced plans to blockade Chinese ports controlled by the Communists, including the bombing of shipping in and around Shanghai. The United States reacted unfavorably, and Webb asked Clark in Guangzhou to tell either Acting President Li or Premier Yan Xishan that the U.S. government would not countenance unprovoked bombing of American vessels and property.[42] The British reaction was even stronger. The Foreign Office cabled the State Department, stating that since aerial bombing must rely on the Chinese Air Force based on Taiwan, it was hoped that the United States would seriously consider cutting off its supplies of aviation fuel. And, if American officers were still on the island training the Chinese flyers, they would be in a position to exercise direct pressure on the units involved in the blockade. The British viewed American presence on Taiwan as support for Nationalist bombing of British ships.[43]

By late June the State Department received word that approximately ten American merchant ships and oil tankers were being threatened by the Nationalists in the Shanghai area. Some were beached, others diverted. Acheson immediately cabled the Chinese Foreign Office to lodge a protest and question the legality of the operation. On June 28 Clark made a personal appeal to Li, asking him to have all bombing stopped.[44] The British were even more upset because on June 21 the British ship *Anchises* was bombed by Nationalist planes outside of Shanghai. Butterworth noted that the British statement of protest was more strongly worded than that of the State Department, for the Foreign Office claimed that the blockade was illegal. Furthermore, it asserted that the Nationalists did not have the means to exercise a real and effective blockade of all of China, and the British were not going to respect the blockade.[45]

The port of Shanghai was effectively closed to foreign trade by mid-July. The British considered this a "matter of urgency" because of the potential danger to communities if public utilities shut down or food riots broke out. In a statement prepared for the State Department, the Foreign Office announced that it was considering sending "relief ships" to evacuate foreign residents and carry supplies to Shanghai. But the State Department did not support this action, especially since it was conjectured that the British were using the idea of an evacuation program as an excuse to reestablish commercial ties with Shanghai. Evidence for this was found in British memoranda explaining that the goals for sending relief ships included an attempt "to keep a foot in the door in China." The British did not want to be forced out of China at this time. There was some evidence, moreover, that British ships had begun running the blockade.[46]

Contention between Great Britain and the United States also arose over the issue of exports to the Communists of military-related material. For six months after the U.S. government approved a restrictive policy on such materials, talks were carried on with the British to convince them to agree to place the same restrictions on their exports, since a unilateral effort by the United States would prove ineffective. The negotiations were unsuccessful, and in July the British announced that they were not prepared to institute a control system over exports.[47] Acheson expressed his disappointment with the decision because such an effort, he felt, was necessary for any united position among Western nations concerning the situation in China. Moreover, without British cooperation, it would also be impossible to receive the support of other exporters to China, such as France, the Netherlands, and Belgium.[48]

Meanwhile, because this policy remained unsettled, Acheson directed Admiral Sidney W. Souers, executive secretary of the NSC, to give clearance to Standard Oil Company of California (also referred to as Standard-Vacuum Oil) and California-Texas Oil Company (Caltex or, sometimes, Texaco), the principal American oil distributors in China and chief competitors of the British-owned Shell Oil Company, to sell kerosene and motor gasoline destined for Chinese Communist-controlled ports. They were to be notified, however, "that it would be desirable to limit such sales to the normal civilian requirements of north China, and that [the State Department] would prefer that, in general, no sales be made for shipment of petroleum to northern Korea," which was considered to be under the control of the Soviet Union. Throughout the summer of 1949, American oil companies sent petroleum tankers to both the Communists and the Nationalists. In both cases there was no control over quantities distributed or determination of its use for military purposes.[49]

In June the Chinese Petroleum Corporation, the Communists' largest oil company, contacted Colonel E. P. Kavanaugh of the California-Texas Oil Company in order to negotiate a three- to five-year contract for the shipment of crude oil to Huludao. While Acheson thought it undesirable to conclude a long-term contract with the Communists at that time, he did not want Kavanaugh to refuse to negotiate since the British-owned Anglo-Iranian Oil Company was also offered such a deal. Caltex wanted to be able to compete for business on equal terms with Anglo-Iranian. Acheson then informed major American oil companies that the State Department wished that any contracts entered into with North China be short term rather than long term. He was also worried that small American concerns might enter the bidding and thereby produce competition that might force large American companies or the Anglo-Iranian Company to extend contract periods in order to procure a deal. The negotiations between Caltex and the Chinese Petroleum Corporation continued for months and were not yet concluded when a message arrived from the president stating that he had heard reports from the CIA that two American tankers were engaged in the transport of petroleum products from Romania to Dalian. Truman wanted to have this traffic stopped, and Admiral Souers suggested that the matter be referred to the State Department.[50]

Truman's statement revealed several problems. The president was obviously uninformed about the tons of oil supplied to the Communists in North China throughout 1949 by American oil companies. These sales, moreover, were legally sanctioned by the president in his approval of the February NSC policy statement. Truman and others in the administration were more concerned with isolating the Chinese Communists than with maintaining economic competition with the British and other Western European nations, the policy being pursued by Acheson and the State Department.

In a memorandum to Acheson giving the details of these developments, Livingston Merchant of the Office of Far Eastern Affairs predicted that there would be "rumblings" at the lower levels of the military establishment and discussions in Congress when this matter was publicized. Merchant suggested that, given the department's policy of supporting the rights of American-owned tankers to compete in trade, Acheson should suggest to the president that no steps be taken to interfere with the tanker traffic until a review of the government's trade policy toward China was completed.[51]

Within a few days the State Department heard from Senator William F. Knowland (R., California), who was seeking information concerning the "petroleum situation" in Communist-occupied areas of China. Assistant Secretary of State Ernest A. Gross prepared a memorandum for Knowland explaining that the

Chinese Communists had over one million barrels from stock of private firms and the Nationalist government when they took over in North China. The most important domestic source of oil was a shale oil refinery in Manchuria. The USSR had been supplying limited quantities of petroleum products to Manchuria and a new trade agreement had recently been signed. North China received shipments from Manchuria, the Soviet Union, and Hong Kong during 1949, but the Nationalist blockade had prevented shipments of petroleum through Shanghai since June. There were indications of a petroleum shortage in China manifested by the recent conversions of the Shanghai Power Plant and other industrial facilities to coal.[52] Thus, Knowland not only did not receive answers to his implied questions, but he also did not learn the reality of American trade in North China.

In the meantime, Kavanaugh of Caltex negotiated a one-year contract on August 26 with the Chinese Petroleum Corporation for crude oil from the Persian Gulf, the first shipment to arrive in Shanghai in January 1950. Kavanaugh met with State Department officials several times to clarify the terms of the negotiations. At one point Sprouse of the Division of Chinese Affairs suggested that Kavanaugh "drag his feet and do nothing toward implementation of this contract" for the time being since there were several problems with which the State Department was then trying to deal.[53] This referred to the public reaction to publication of a policy statement known as China White Paper and the apparent "abandonment" of the Nationalist government. Knowland and others accused the department of setting up a "colossal fraud," claiming that the U.S. government had not given any aid at all to the Nationalists. Acheson has stated that it was this situation that prevented him from giving serious thought to the question of recognition at that time.[54]

Conclusion

In the summer of 1949 the Truman administration pursued several contradictory policies toward China. Millions of dollars of aid and material were shipped to Taiwan to aid the Nationalists who were setting up a regime there while relations with the CCP were maintained in order to keep a toe-hold in the mainland. American oil companies sold petroleum products to both the Nationalists and the Communists. The president gave verbal support to a Nationalist blockade of the mainland, while the State Department permitted American firms to negotiate contracts with the Chinese Communists. Attempts continued to win the release of American hostages in Mukden. In August the China White Paper was released to clarify the American position in China.

5. An Inconsistent Policy

By late 1949, the Truman administration had to deal with the inevitable victory of the Communists on the mainland and the Nationalist retreat to Taiwan. U.S. policy makers balanced support for the Nationalists with a desire to contend with the question of recognition for the new Chinese government. Chiang Kai-shek's supporters were pressuring the administration to redouble efforts to prop up the failing Nationalist government. Truman dealt with these issues by apparently divorcing the United States from the outcome of events in China. In August 1949 a report, the China White Paper, announced that the United States would stand back from further involvement.

The policy that emerged was actually an attempt to play both sides—to appear to be open to the notion of recognition and to protect Taiwan. Such a policy did not sit well with either the Communists, who still held American hostages at Mukden, or the Nationalists on Taiwan. Furthermore, any semblance of solidarity among Western nations was undermined by several European allies, particularly the British and the French, who paved the way for recognition by late 1949. The Americans seemed to be isolated.

Development and Release of the White Paper

On August 5, 1949, the State Department released a publication entitled *United States Relations with China, With Special Refer-*

ence to the Period 1944-1949, often referred to as the China White Paper. It announced an American plan to "abandon" the Nationalist Chinese government since nothing could be done to prevent a Communist victory. Its publication coincided with negative responses by the Truman administration to requests for more money by Nationalists remaining on the mainland under President Li Zongren. The document, based on State Department files, summarized relations between China and the United States since 1844, but it gave more detailed coverage to the period during and after World War II. According to Secretary of State Acheson in his "Letter of Transmittal," which served as an introduction to the volume, the study revealed the "salient facts" determining United States policy toward China.[1]

The release of the White Paper was accompanied by a news release by President Truman on August 5 and a press conference by Secretary Acheson on the following day explaining what was done, how it was done, and who had done it in order to "make it understandable for the press."[2] Detailed documentation of U.S. aid to the Nationalists from 1937, American efforts at bolstering Nationalist military strength, the Marshall mission and other attempts at Nationalist-Communist negotiations, and the corruption, incompetence, and ineptness of the Nationalist regime were presented to explain why it was "the unfortunate but inescapable fact . . . that the ominous result of the civil war in China was beyond the control of the government of the United States." Acheson stressed that it was "abundantly clear that we must face the situation as it exists in fact. We will not help the Chinese or ourselves by basing our policy on wishful thinking."[3]

By releasing this document before the total collapse of the Nationalist government on the mainland, the Truman administration accomplished several things: It divorced itself from identification with the Nationalists and their eventual and inevitable defeat and answered criticism in the United States that American policy was wavering and inconsistent. It pushed the Nationalists to become more independent of the United States by asserting that the Americans would not defend a Nationalist government on Taiwan. In this way, it was hoped that the Truman administration would not have to make a military commitment to the island. On the other hand, the document was an attempt to open the door to negotiations with the CCP on the question of recognition and, perhaps, make it easier to remedy the Mukden hostage situation.

The White Paper was the brainchild of W. Walton Butterworth, director of the Office of Far Eastern Affairs. His original idea was to lay out the objective story of U.S. policy toward China in the postwar period. It was to be a document that would stand up historically and would be published well after the Nationalists' defeat and the Communists' takeover. Butterworth issued specific instructions that it should hide nothing and was not to be a partisan apologia for the Democratic administration.[4]

Work was under way on the project by April 1949. Although there was no formal directive for the preparation of the White Paper, clearance was approved by Acheson after consultation with Truman. The president acknowledged his support for the project, which was given top priority and was to be completed within several weeks. Truman took considerable interest in the preparation of the paper, reading and commenting on the various sections and drafts and noting whether certain segments should be included or omitted.[5]

The compilation of relevant documents by the State Department continued throughout the spring. Acheson has remarked that it was not until his return from a foreign ministers' meeting in Paris on June 23 that he fully perceived the potential problems resulting from such a project. By then a huge quantity of material had been collected and something had to be done to trim it to a manageable size. There were also several puzzling questions. What was the basis for selection of these materials? Was this the proper basis? Who did the selecting? Was it done by anyone who was trying to defend his own actions? Did someone selectively try to pick the materials? Acheson stated: "I dealt with [these problems] in a way which was most unkind to one of closest friends, which was by asking [Ambassador] Phil[ip C. Jessup] to take charge of it, and be the editor, to assure himself that he had all the material that was relevant, to go over it himself, to be sure that nothing was improperly done here, and to put it in some kind of a form which brought order out of it so that anyone who wanted to read it could at least follow a theme here."[6]

Acheson then had to consider whether and when the paper should be released. He discussed this with the president, and it was decided that if Acheson "was satisfied that this document was a thorough, honest, scholarly document, it should be brought out."[7] The question of when it should be released was not solved so easily.

By July Jessup was working on the paper with four teams under his direction. As the preliminary drafts were prepared, members of several branches of the government became involved in the critical reading. For example, the president had asked that former Secretary of State George C. Marshall read it. Clark M. Clifford, special counsel to the president, prepared a lengthy memorandum on July 5 outlining what he considered to be serious gaps and poorly written segments in the text. The secretary of defense and the Joint Chiefs of Staff were also among those who commented on portions of the document.[8]

Although the readers accepted the concept of a white paper on U.S. relations with China, many did not agree with an early date for its publication. "Butterworth," Acheson remembered, "was very insistent on postponing it until the last flicker of hope [for the Nationalists] went out."[9] Butterworth also predicted that the issuing of the paper would seriously jeopardize any appearance of a bipartisan foreign policy toward China as embodied in earlier aid acts supported by Arthur H. Vandenberg, ranking Republican member of the Senate Foreign Relations Committee. Marshall commented that publication of the report while there was any possibility of resistance still left on the mainland would minimize such resistance. The Joint Chiefs strongly objected to early publication for security reasons. Moreover, after the news of the study's preparation reached Nationalist leaders, there was "increasing perturbation in [Chinese] government circles over [the] possibility [of its] release."[10]

It is apparent that Truman made the final decision concerning the White Paper's release. The president followed the progress of the volume's preparation, noting, for example, that he was pleased with Clark Clifford's report that an edited draft was "an excellent paper." As for criticism by the Joint Chiefs, Truman assured Acheson that he would give him "all possible assistance in dealing with the Military" because he did not want the paper to be "watered down."[11]

On July 21 Acheson spoke with Truman on the subject of postponing the publication date. He informed the president of Jessup's decision made on the preceding day to revise the schedule and push back the publication date at least a week because of the strong objections of the military establishment. The president, however, was of the opinion that the schedule should continue as planned. Acheson and Truman met again on July 25 to discuss the same question. Although Acheson noted that Secretary of Defense Louis Johnson concurred with the Joint Chiefs, the president again said that he believed it was "necessary and desirable to bring out the White Paper."[12]

On July 29 Acheson gave Truman a memorandum outlining the questions raised by opponents of the White Paper's publication. By this time the final revisions and corrections in the text had been made and the copy had gone to the Government Printing Office. Truman decided that the White Paper would be released as soon as possible; he planned to review the final copy during the next weekend. In reply to the criticisms of the military establishment, Acheson drafted a memorandum to the secretary of defense on August 3 stating that he had informed the president of the opposing views and, after careful reconsideration, the president decided that the paper should be issued on August 5.[13]

According to Acheson, he and Truman "realized that this was going to cause quite a lot of trouble" and they "spent a good deal of time on thinking what [to do] next." They devised a plan whereby two Republican consultants would be brought in to work with the State Department. They "thought this would give an appearance—not an appearance but the actuality—of bi-partisanship."[14]

The timing of the paper's release was important. Philip D. Sprouse, then director of the Office of Chinese Affairs, has commented on this subject:

> At one point, in a meeting in the Secretary's office, when Dean Acheson was Secretary of State, if I remember correctly, this [the White Paper's release] began to get into the political mill. And for political reasons, I'm convinced—I can't prove this—it was decided to publish the white paper earlier rather than later. I think one of the reasons it was done, was that the situation in China was being used in a very partisan way by the Republicans to belabor the Democratic administration and to charge them with responsibility for the loss of China to the Chinese Communists. So the Chinese white paper was published earlier than it was expected to be when we were writing it, and earlier, I think, than it should have been published. This was a tremendous shock, I'm sure, to the opposition.[15]

But this shock wore off quickly. Congressional supporters of the Nationalists reacted by calling the volume a "whitewash." Representative Walter H. Judd (R., Minnesota) and Senator Henry Cabot Lodge, Jr. (R., Massachusetts) retaliated by proposing additional funds for China as part of a broad military assistance bill being debated in the House of Representatives. Acheson told several members of the House Foreign Affairs Committee that he opposed such a proposal.[16]

The White Paper and Taiwan

The State Department's strategy of apparently divorcing itself from the fate of the Nationalists is relevant to the situation on Taiwan at the time. It supported the department's goals for the island in two ways. First, because the American commitment to the island so far had fallen short of utilizing military force to keep it from the Communists, the White Paper's release was designed to notify the Nationalists that they themselves were responsible for the security of the island. An increase in the amount or kind of assistance should not be expected, and seeking additional assistance with the help of the China bloc in Congress would be useless. It was hoped that this policy would establish an anti-Communist government on the island that did not require American military strength to maintain itself.

This part of the strategy was somewhat successful. One of the first official reactions to the White Paper's publication came from Taiwan's Governor Chen Cheng. In an address to an agricultural meeting in Taibei on August 8 he said that the White Paper "should in no way dampen [the] hearts of Chinese in resisting Communists. Instead, it should be a blessing to [the] Chinese people. In [the] past, people have been relying too much on support for agricultural help from [the] U.S. This had created subconsciously in minds and in attitude [a] deep spirit of dependence, which is subjecting [the] nation to [the] status of semi-colony. [P]ublication of [the] White Paper should be on our own entirely and self help is the best help."[17] Acting President Li Zongren also agreed that the "fundamental" problems in China must be solved by the Chinese and not by American bullets. In a meeting with Clifford in Canton on August 10, Li said that he would try to carry out needed reforms in his mainland government but would continue to seek American aid.[18]

The second way in which the White Paper's publication supported the State Department's goals for Taiwan was to demonstrate, through the public "rejection" of the Nationalist government, that the U.S. government was not associated with that group on Taiwan and thereby dismiss accusations made by the Soviets and the Chinese Communists that the United States had designs on the island. This was to be a step, albeit minor, toward reaching an accommodation with the new government in China once it was officially established. The Communist reaction, however, as seen in the publications *Jiefang ribao* and *Xin Hua*, was just the opposite. Despite precautions of secrecy surrounding the developing American policy toward Taiwan during 1949, the Chinese Communist press made it clear that, even if the American public was sure that its government had abandoned the

Nationalists on Taiwan, the Communists correctly realized that American aid continued to pour onto the island. For the Communists, the White Paper "reveal[ed the] ugly face [of] American imperialism in its attempt [to] attain world hegemony and [its] self-appointed role [as] protector of China."[19]

The publication of the White Paper was consistent with other U.S. attempts to approach the Communists indirectly. For example, on June 6 Ambassador Stuart met with Luo Haisha, a leader in the Nationalist Party's Revolutionary Committee, a group which did not support Chiang. Luo was on his way to Beijing to meet with high-level Communist Party leaders. He asked Stuart what he should say to the Communists when they asked why aid was still going to Taiwan if the United States did not have designs on the island. Stuart replied "that perhaps [the] simplest thing would be to ask CCP leaders to look at [the] facts. We could have included Taiwan in SCAP [Supreme Command, Allied Powers in Japan] or even claimed it as our share of postwar settlement as USSR did with Sakhalin and Kuriles, to say nothing of Manchuria."[20]

In reality the White Paper did not reflect a change in policy. The State Department continued to monitor the situation in Taiwan. On the day preceding the White Paper's release, August 4, the department sent a memorandum to Souers, the executive secretary of the National Security Council, on the subject of the "Current Position of the U.S. with Respect to Formosa."[21] Developments on the island during the summer made it desirable to reexamine the policy set forth in February, which was summarized as follows:

> To attain our main objective with respect for Formosa and the Pescadores—the denial of the islands to Communist control—current policy directives . . . call for (1) developing and supporting a local non-Communist Chinese regime which will provide at least a modicum of decent government for the islands; (2) discouraging the further influx of mainland Chinese; and (3) maintaining discreet contact with potential native For-

mosan leaders in the event that some future use of a Formosan autonomous movement should be in the United States national interest.

The memorandum gave several reasons for reconsideration of this policy. First, the mass movement of troops and civilian refugees from the mainland was causing serious economic damage to the island, including runaway inflation. Second, after peace negotiations between the Nationalists on the mainland and the Communists had broken down, there was little possibility of the island's being a bargaining point in a movement for a coalition government, thus eliminating the chance to use a local Taiwanese independence group and the United Nations to secure the island. Third, the Li-Chiang feud served to divide the Nationalists, with Chiang and his supporters firmly establishing themselves on Taiwan and, as a result, the hope for the installation of an effective and liberal government declining. In view of this situation, it was decided that ECA funds would be used only for the importation of consumer goods, principally fertilizer and cotton, and for rural development programs rather than for capital reconstruction projects that would enhance the inflationary trend.

The State Department memorandum assessed the U.S. position in this way:

> We face on Formosa today a situation analogous to that which confronted us on the mainland of China a year ago. The government in power is corrupt and incompetent. It lacks the will to take the necessary political and economic steps to modify the deep and growing resentment of the Formosans. The burden of supporting the mass of Nationalist troops and governmental establishments is now so great as to accelerate the economic disintegration of the island. Moreover, economic aid from our side cannot in the absence of a basic change in the government alter or cure this situation, and so long as it endures the ultimate passage of Formosa under Communist control, by external or internal action, appears probable. The Governor of Formosa has reiterated his intention to resist any Communist

assault on the island. The forces on the island appear numerically sufficient provided they will fight. It is also believed that the Governor now controls sufficient troops to suppress any native insurrection. The most serious risk of a turnover of the island to the Communists lies in the possibility of widespread mutinies by disaffected Nationalist troops recently landed from the mainland and estimated now to have brought the total to the neighborhood of 300,000, or in deals with the Communists on the part of the top military commanders. Such a development could occur at any time and would confront us with a *fait accompli* which only military force could reverse.

As a result of these developments, there were "no certain assurances that [Taiwan could] be denied to Communist control by political and economic measures alone." Therefore, the department requested that the Joint Chiefs of Staff, "as a matter of priority," advise the National Security Council on U.S. military interest in Taiwan, taking into consideration two types of situations: (a) a U.S. military occupation in the face of initial opposition by Nationalist forces on the island or an attack from the mainland by Communist forces, and (b) an occupation in cooperation with the island's authorities with the responsibility for internal and external security falling upon the United States. In the meantime, the State Department planned to continue to exert its influence on the governor of Taiwan to assure that he maintained his will to resist Communist influence while he adopted constructive political and economic programs designed to alleviate the problems among the population that left them susceptible to Communist propaganda. The ECA program would be continued with a concentration on the importation of consumer goods. Finally, informal talks would begin with "selected governments, particularly the British, with a view to securing their views and laying the ground work for possible future joint or concerted action within or without the framework of the United Nations."

The State Department received the Joint Chiefs' reply on August 22.[22] In the updated position concerning Taiwan, the earlier emphasis on the strategic importance of the island was not changed. Moreover, since the February policy statement was developed, the continuation of the Communist sweep across the mainland only strengthened this view. On the other hand, the Joint Chiefs reaffirmed their view that the island's strategic importance "does not justify overt military action, in the event that diplomatic and economic steps prove unsuccessful to prevent Communist domination, so long as the present disparity between our military strength and our global obligations exist, a disparity that may well increase as a result of budgetary limitations and the commitments implicit in the North Atlantic Treaty.

The suggestion of an American military occupation with respect to the two possible variations in the situation on Taiwan was also addressed. The Joint Chiefs concluded that an occupation in the face of opposition from Nationalist forces or an attack by mainland Communists would lead to the need for a "relatively major effort" that could make it impossible for the United States to meet emergencies elsewhere. An occupation with the consent of existing authorities, although less embarrassing than the first alternative, might also lead to a long-term American military commitment which was undesirable at that time.

But this strong feeling against an American military commitment to Taiwan was modified in a concluding paragraph, where it was explained that this attitude might change at some point in the future:

> Although the Joint Chiefs of Staff are of the opinion, on balance, that if Communist domination of Formosa cannot be denied by diplomatic and economic steps, military measures instead of or in support of diplomatic and economic efforts would be unwise, they must point out, as they have previously stated, that future circumstances, extending to war itself, might make overt military action with respect to Formosa eventually advisable from the over-all standpoint of national security. They believe that it is better, however, to face this future contingency as one of the many military problems that must be considered in the event of incipient or actual overt war than to

risk undue military commitment in the Formosan area under present circumstances.

This policy statement reflected the Joint Chiefs' position during August and September 1949. Meanwhile, reports from the consul general at Taibei, John J. MacDonald, were received describing a deteriorating local situation, manifested by almost daily public executions, banditry, unsafe conditions for foreigners, and a lack of control over military personnel. MacDonald stated that the Taiwanese were predicting trouble. Feelings against mainland Chinese were more intense than in February 1947, and a second bloodbath would be even greater.

Also at this time, the State Department carried on talks with the British on such ideas as an independent Taiwan and use of the United Nations to work on the problem. The British, however, were less than optimistic about the prospects for maintenance of a non-Communist Taiwan if the United States did not act militarily.[23]

Establishment of the PRC

The declaration of the founding of the People's Republic of China on October 1, 1949, forced the State Department to reappraise American interests in China once again. On October 6 the National Security Council prepared a study at the request of the secretary of state to determine what modifications should be made in existing policy in light of the fact that the Communists now held de facto control over most of the mainland. It was decided to follow essentially the same policy outlined in earlier statements with a few important changes. Because Chiang Kai-shek was now the "real source of authority" on Taiwan, many of the administrative weaknesses on the island were a result of problems that had plagued the Nationalists on the mainland and had been transferred to the island. The State Department was aware through its own intelligence and through Philippine sources that Chiang claimed privately that he had sufficient resources to defend Taiwan for at least two years without outside assistance. Also, the major portion of the military material purchased from the United States under the $125 million grant was stockpiled in Taiwan. As a result, Washington decided to let Chiang know that it was primarily his responsibility to defend the island, although the United States was "concerned . . . lest the chaos of the mainland spread" to the island. Americans and Chinese were to be told that Taiwan was the sole responsibility of the Chinese, but further consideration was given to the use of the United Nations as a possible arena for action if the Communists decided to take the island.[24]

In a letter dated November 18, Acheson outlined the policy that MacDonald in Taibei should pursue in his relations with the Nationalists.

> U.S. policy is [to] endeavor [to] deny Formosa [to] Commies by using polit[ical] and eco[omic] means. Resources on Island now available [to] Chi[nese] appear adequate at this time if resolute steps [are] taken [to] utilize those resources effectively; until these steps [are] taken, any commitment of increased U.S. support w[ou]ld not contribute to achieving our objective since it w[ou]ld probably lead to Chi[nese] conviction [that] U.S. [is] assuming role and responsibility and that Chi[nese] self-help measures [are] not essential and [this] w[ou]ld thus in [the] end be costly to U.S. in terms of its prestige as well as resources and very possibly [to] achievement [of] political objectives on [the] mainland.[25]

The State Department's political objectives for the mainland were not yet clearly spelled out, at least concerning the recognition of the new regime, for several reasons. The Nationalists under Li Zongren were still there, and it was this government that the United States officially recognized. This was demonstrated on the Chinese national holiday, October 10, 1949, when President Truman sent a telegram to Li expressing good wishes to him and to the people of China.[26] Furthermore, some areas south of the Yangzi River remained under Nationalist control after October 1, and another possible option for the State Department was to survey the possibility of using these areas as anti-Communist strongholds. Rejection of the various requests and proposals supported the policy of abandoning the mainland and adopting a "wait-and-see" attitude toward the Communists.

Aid to Yunnan

One area that appeared promising as a pocket of resistance to Communist expansion was Yunnan, far to the southwest. Despite Marshall's statement that such resistance would be effectively undermined by the release of the White Paper, considerable activity was carried on for several months after the founding of the People's Republic. Yunnan, with its high elevation and mountainous terrain, was to be the site of a "last-ditch stand" by those Nationalists who did not retreat to Taiwan during the summer with Chiang Kai-shek. On October 19 Chu Changwei, political adviser to Li, announced that Kunming, Yunnan's capital, would be the home of the Executive Yuan as soon as it was possible for the government to remain in Chongqing. Kunming was a particularly appropriate location because of its history as a Nationalist retreat during World War II as an anti-Japanese base. It was, for example, the site of three northern universities during the war. Its geographical remoteness plus its highly developed civil air transport made it a desirable site. Chu admitted that the Communists had the strength to capture Sichuan, east of Yunnan, but the government would hold out for a long time in Yunnan. In early 1949 the Nationalist 26th and 49th armies had been transferred to Yunnan, and the Provincial Peace Preservation Corps had been expanded from six to twelve regiments. A secret arms deal had been negotiated with a French firm, Rondon et Cie, to supply these forces without direct aid from the Nationalist treasury under Chiang's control.[27]

In early November the State Department was faced with a new problem concerning this area. While the mainland Nationalist group was preparing to set up a government in Kunming, a group claiming to represent top-ranking local government officials and businessmen approached the American vice consul there, LaRue R. Lutkins, promising to "do anything desired by [the] U.S. Government" if "America issue[d a] statement promising to defend Yunnanese independence and territorial integrity."[28] In its request the group referred to Truman's statement made on December 30, 1948, that the U.S. government would aid any group in China that actively opposed communism. The group emphasized the strategic importance of Yunnan as a buffer against Communist expansion in China and Southeast Asia.

This was particularly important given the fact that beginning in October Chu Changwei continuously complained of the difficulty in containing Communist forces in the South because of aid the Chinese Communist forces were allegedly receiving from the Vietnamese Communists. Chu had stated that Vietnamese Communist forces had crossed into Guangxi on orders of their leader, Ho Chi Minh, in response to requests by the Chinese Communists. Chu had been authorized, he said, to ask the State Department to use its influence with the French to allow Chinese Nationalist armies to carry out punitive missions in northern Vietnam and to remain there as a precautionary measure. The American chargé in Chongqing, Robert C. Strong, initially considered this information "rather amazing" and designed to give legal sanction to a southern escape route for Nationalist forces, but Lutkins in Kunming soon confirmed the claims, noting that both French and Chinese sources agreed that there was some cooperation between the Yunnanese and Vietnamese Communists. According to these sources, some Yunnanese Communists received training across the border in Hagiang, orders had been sent through Hagiang radio, and a small quantity of arms and ammunition had been shipped from Vietnam for the Chinese.[29]

The State Department did not look favorably on the Nationalists' request. Li was informed that, since the Chinese government maintained diplomatic relations with the French and there were French representatives in Chongqing and Chinese representatives in Paris, the two governments should discuss the matter directly without the involvement of a third power. To Strong, the Nationalists' motive was to involve the United States somehow in a serious international incident by encroaching upon French territory, something that the White Paper had warned the Communists not to do.[30] Strong concluded that many Nationalist leaders felt that their China would be saved only by a third world war. He wrote to the department on November 10 that "Since time is running out on them, they may well try to assist in development in international stituation before Chinese Government is defunct on mainland. . . . We feel that men who consider it practically a right to receive large-scale aid from U.S. and who are thoroughly disappointed in this respect may well seek to force trouble on U.S. They have no future in Communist China nor do they have a future outside it."[31]

The State Department did not give serious consideration to the notion of aid to Yunnan as an independent state or to the remains of the Nationalist government in South China, despite Acheson's assertion to the Senate Foreign Relations Executive Committee on October 12 that Yunnan along with Guangxi and Hong Kong were still valuable as footholds in China. There were several reasons for the futility of such an endeavor. As early as June 1949 the CIA had completed a study on the use of South China as an anti-Communist base and had concluded that there were "seemingly insuperable obstacles" standing in the way of the establishment of that area as a Nationalist stronghold. The problems included insufficient time to develop necessary industrial and military installations, inadequate food supplies for large numbers of refugees, disunity among the Chinese there, potential hostility toward the transplanted Nationalists, and a lack of capital.[32]

Lu Han, governor of Yunnan, also expressed pessimism concerning the possibility of effective resistance from the province unless an American military mission were sent to the area. The Communists would not be halted by the mountainous southwestern terrain as the Japanese had been during World War II because of the apathy of the Chinese people, who did not consider the Communists as invaders. Lu himself was of questionable loyalty to the Nationalist cause because of reports received in August by the Chinese Bureau of Investigation and Statistics (Security) that he was preparing to go over to the Communists. Until mid-August, he had been "fence-sitting," waiting to see which group would emerge as the obvious victor.[33]

Lutkins agreed with Lu's assessment. In early November he cabled the secretary of state with his evaluation of the military situation in Yunnan: "Like most observers here, I feel [that the] Commies can take Yunnan when they wish; time depends only on their plans and food supply problems. I am convinced that mountainous terrain will offer no effective obstacle. Province may go within 2 months and there is little hope of effective Nationalist resistance here for more than 6."[34]

In a telegram dated November 22 the State Department dismissed the idea of sending military aid to defend Yunnan. It explained to Lutkins why there were no grounds for assuming that military assistance, even in amounts requested, would enable the Yunnanese Nationalists to maintain themselves against overwhelming Communist strength. The Joint Chiefs of Staff considered that the possibility of the development of a successful resistance movement was too remote to justify American involvement. Furthermore, it could be assumed that the Communists would "almost inevitably" be the ultimate recipients of material provided. As for a statement guaranteeing Yunnanese independence, this was also considered impossible by the State Department. Among the reasons given was that it would encourage the breakup of the Nationalist government and constitute "intervention [in] internal affairs [of] China." Nonetheless, the Yunnanese independence group resisted, declaring on December 6 that they would proclaim independence with or without American approval. Their chief concern was finances, and they asked if the United States were willing to purchase vast quantities of Yunnanese opium for "medical" use. They were told that this was unlikely and that there would be no probable recognition for Yunnan by the United States if independence were declared. Independence was not proclaimed.[35]

Focus on the Mukden Problem

Although the State Department would not consider supporting an independent Yunnan, the question of recognition of a Communist Chinese government became more deeply intertwined with the Mukden hostage situation by the end of 1949. On October 1 Zhou Enlai sent an official communication to Clubb announcing the founding of the People's Republic of China and stating that it was necessary to establish normal diplomatic relations between the PRC and all countries. In response Clubb despatched a lengthy letter to People's Liberation Army Commander in Chief Zhu De criticizing the "arbitrary and unreasonable restriction" placed on Americans in Mukden and asking Zhu to take appropriate measures to cause Mukden authorities to deal with the matter promptly. On October 4 Acting Secretary of State Webb asked Clubb to send a similar communication to Zhou Enlai, now premier and foreign minister, emphasizing in particular the approaching winter season, which would impose extreme hardship on the Americans in Mukden under present conditions.[36]

During the following week, an effort was made to coordinate a position among the Western allies, but these attempts failed. Despite what members of the State Department felt to be a policy agreed upon by many Western allies, including Great Britain and France, to keep each other informed of moves on the question of recognition, the British sent a reply to Zhou on October 6 that in effect accorded de facto recognition to the new government.[37] But what was more upsetting for Acheson was the idea that the Foreign Office did not inform the State Department before sending such a communication, and further, that he had received a copy of the British reply not from the British but from the French embassy, which was notified before the State Department. When Acheson brought up this issue with the president at a meeting on October 17, Truman noted that the Soviet reaction to the British note was that it amounted to recognition. He remarked that "the British had not played very squarely with us on this matter."[38]

In its reply to Zhou's bid for recognition on October 10, the

State Department merely acknowledged receipt of his letter of October 1. The remainder of the message dealt with a description of the situation in Mukden and an appeal to Zhou to allow the Americans there to depart promptly for the United States. In conclusion, it stated that the "United States Government is deeply concerned with this situation, which is contrary to established principles of international comity and which has been permitted to continue despite representations to the Chinese Communist Military Headquarters."[39] On the same day, the president sent his congratulatory message to Acting Nationalist President Li Zongren on the anniversary of the 1911 Revolution.

By this time it was clear that any State Department attempt to isolate the Chinese Communists diplomatically or economically would be difficult, if not impossible. The USSR recognized the PRC on October 2, and within a few days the Eastern European states followed. Whereas this was to be expected, it was also obvious that many Western European and Asian states were merely delaying the recognition until one of the "major powers" recognized the new government first. Meanwhile, the State Department continued to cable its embassies throughout the world asking ambassadors to convince their host governments not to be too hasty in according recognition.[40]

On October 16 another communication was received from Mukden which only served to exacerbate tensions within the State Department. Ward related the details of an incident at the American consulate where he and four staff members were accused of beating two Chinese employees. Ward claimed two Chinese laborers had feigned injuries but an attending physician had announced that one had a cerebral concussion. News releases from Mukden accused Ward of leading a mob in beating the workers. On October 27 Ward and the accused staff members were "removed" from the consulate for two or three days. Upon receiving this information, Acheson cabled Clubb and told him to ask for an appointment immediately with Zhou Enlai to tell him that the U.S. government considered this situation with the "greatest concern." Clubb was also to ask him whether the continued failure of Communist authorities to facilitate the withdrawal of the Americans was to be considered a manifestation of hostility toward the United States and its nationals in China. He was to conclude by stating that the U.S. government fully expected the highest Chinese Communist authorities to take appropriate measures to allow the Americans in Mukden to leave immediately.[41]

On the next day Acheson received word that a petition signed by thirty-five workers and employees had been filed against Ward charging him with violating his employees' human rights. In accordance with judicial procedure, the Shenyang [Mukden] Municipal People's Court had formally accepted the case, and Ward and the other accused staff members had been placed under detention. In response to this development and to his lack of success in arranging a meeting with Zhou, Clubb dispatched a letter to Zhou reiterating the various requests already made by the State Department concerning the situation and informing Zhou that the U.S. government had decided to close several consulates in North China, including that at Mukden.[42]

Alternative courses for action were also considered. At a meeting on October 31, Truman suggested to Acting Secretary of State Webb that he was prepared to take the "strongest possible measures" including using some force if necessary. He wondered if it would be possible to send in a plane to take the Americans out of Mukden.[43] This suggestion was not well received in the State Department. Assistant Secretary of State for Far Eastern Affairs Butterworth pointed out that such a move would entail the risk of spreading hostilities toward other Americans in China as well as the immediate threat to those in Mukden. Butterworth informed Webb that if he were to discuss this matter with the president again, he should outline some of the "less drastic" measures the department was considering, such as appealing to high-level Communist authorities, to the United Nations, or to the Soviet Union. To this end, during the following weeks Webb asked Americans at the consulates in Hong Kong, Nanjing, Beijing, Shanghai, and Tianjin to speak with Chinese officials concerning the Ward case. Ambassador Stuart, the former president of Yanjing University in Beijing, cabled several of his old associates who were now CCP members. Acheson sent a lengthy letter to the foreign ministers of thirty nations asking them to lodge a protest with Chinese authorities over the treatment of the Americans in Mukden. Half of the countries agreed to protest, the notable exceptions being the USSR and two notes of sympathy without protest by Great Britain and Portugal.[44]

The Mukden Situation Is Resolved

In November in Mukden a trial of the Americans and their staff involved in the alleged beatings of Chinese employees took place. Ward cabled that he had been refused legal counsel, the right to produce witnesses for defense, and the right to question witnesses or rebut arguments. On November 28, 1949, the verdict was made public. Ward, as principal defendant, was found guilty and given a six-month suspended sentence. The staff members—two Americans, one German, and one Italian—were given suspended sentences of three to four months. Ward was to pay the principal plaintiff three months' back pay and two months' severance pay, and retirement deductions were to be refunded. The five were to be deported from Manchuria.[45]

The last part of the sentence was good news for the State Department, although Mukden authorities did not move quickly on its implementation. But other upsetting news was heard from Mukden. The five prisoners had been kept in solitary confinement for over one month on six slices of bread and three ounces of hot water a day. Ward had lost twenty-five pounds and was in poor physical condition along with two of the others. The other American, Shiro Tatsumi, was suffering from shock and mental strain and was incapable of rational thinking or speech. And yet another problem had arisen. Another American at Mukden, Vice Consul William N. Stokes, was taken from the consulate without a warrant on November 26 for a hearing concerning "spying charges." On the following day it was reported that some 1,000 people attended the hearing on an "American spy ring." Stokes was merely an observer. At the trial's end it was announced that all non-Chinese staff of the consulate were to be deported.⁴⁶

In light of these developments Webb asked Clubb in Beijing to send a letter immediately to Zhou Enlai asking that facilities be made available for travel for all the consulate staff in Mukden. Acheson also asked Clubb to ascertain whether he could obtain medical treatment in Mukden for the consulate staff. If necessary, he was to enlist the aid of the British consulate there.⁴⁷

On December 2 the Americans in Mukden were given their deportation notices; they were to be out of the city by December 5. Communist authorities delayed, however, in making travel arrangements, but finally on December 10 the Americans and other foreigners departed Mukden.⁴⁸

The resolution of this conflict between the Communist authorities in Mukden and the U.S. government theoretically paved the way for the possible recognition of the PRC as outlined in Acheson's policy statement of May 13, 1949. By the time the Mukden staff was traveling home, it was clear that, with few exceptions, most governments friendly to the United States would soon recognize the PRC. Furthermore, many of the smaller European states were merely waiting for a bigger country, such as Great Britain, to take the lead.⁴⁹ On December 16 Acheson received a personal message from Bevin informing him that the British cabinet had decided to recognize the Chinese Communist government during the first week in January. On December 24, the State Department was informed that the date would be January 6.⁵⁰

The Truman administration adopted a "wait-and-see" attitude toward the question of recognition. During a late December cabinet meeting, Acheson said that the State Department held the firm belief that "our people will be treated respectfully or there is no point in recognizing them."⁵¹ In a December 20 meeting with the president, Acheson reported the decision of various governments regarding this issue. He told Truman that the State Department would make no recommendations to him until "the matter developed further" and there was time for consultations with Congress. The State Department was aware that members of Congress had been inundated with letters and telegrams from Roman Catholic organizations and individuals as well as other groups opposed to recognition of the new government. The State Department itself had also received "piles" of similar correspondence along with many requests from various representatives asking for information with which to reply to constituents. Truman approved the decision to wait and, according to Acheson, was apparently undisturbed by the knowledge that other governments might take action soon.[52]

6. American Interests in China Are Jeopardized

Months before the Mukden hostage situation had been resolved, the Truman administration was faced with other problems for U.S. interests in China. The issue of recognition of the PRC and the importance of Taiwan were tied to concerns over American business interests in China and economic competition with the British. Secretary of State Dean Acheson commented that he did not want to see the British picking up the trade abandoned by Americans. The Foreign Office had responded to pressure from British business interests in China with a statement on November 1, 1949, announcing that Britain "advocated a policy of keeping a foot in the door." To do so, recognition of the Chinese Communist government was necessary. Other U.S. allies, including the French, agreed, and the Americans began to feel increasingly isolated in their attempts to deal with the release of the Americans in Mukden. The Truman administration did not necessarily want to represent the only major power that refused to establish diplomatic relations. As a result, U.S. policies were often contradictory.

Although recognition was not forthcoming, business with the Communists continued until late 1950, after the outbreak of the Korean War and after American consulates in China were closed and their staffs withdrawn. This situation led to tensions on several sides. As the British made overtures to the Chinese Communists, the Chinese favored the British to the Americans' detriment. The Chinese were sensitive to the fact that the U.S. government not only continued to recognize Chiang Kai-shek's government on Taiwan but was giving him military and economic support. The latter also angered the British, who felt that the Nationalists were using American military equipment to bomb British holdings on the mainland. Truman's "wait-and-see" policy jeopardized American interests in China.

Truman Supports the Nationalist Blockade

The blockade of northern Chinese ports set up by the Nationalists on Taiwan in June 1949 to prevent foreign access to Communist-controlled areas posed problems for international business interests in China. Although it was the official American policy to consider such ports closed, several American-owned merchant vessels tried to run the blockade in order to reach Shanghai and Qingdao. Acheson wanted to clarify that the State Department did not support such actions and that the decision to undertake them rested with the shipping companies.[1]

When Truman received the information concerning American trade in China, he ordered Acheson to begin a revision of the February 1949 National Security Council policy that allowed continued trade. Moreover, after hearing that three of the American vessels attempting to run the blockade had been captured and detained by Nationalist warships, Truman "indicated strongly . . . that he wished that blockade to be effective and that he desired the Department to do nothing to be of assistance to these vessels." He stressed that "his policy was to permit the blockade to work effectively to which policy he expected strict adherence." This statement was made on the day of the founding of the PRC, at the same time as the State Department was trying in vain to contact Zhou Enlai concerning the conditions in Mukden. Five days later the department heard through the French that the British had acknowledged de facto recognition of the new authorities. Within a week, on October 10, the president received a message that the British Royal Navy would escort British vessels bound for Shanghai and territorial waters at the mouth of the Yangzi River. Truman expressed his surprise at what he considered to be "a most undesirable development." Acheson initially doubted the report's validity because of his earlier agreement with Bevin in which each had promised to keep in close touch with the other concerning actions in China. But in his inquiry to the Foreign Office Acheson learned that the British government had indeed authorized its navy to protect British ships up to the limits of Chinese territorial waters (three miles), which included the areas where ships had been seized. Once these ships entered Chinese territorial waters, they would not be escorted unless attacked or bombed. In that case, the navy would enter to protect them. If the ships were arrested and forced back, then nothing would be done unless the vessels were forced to go to a Nationalist base. If so, the British navy would intervene to stop this.[2]

The owner of the three American vessels detained in Shanghai waters in October, the Isbrandtsen Company of New York, cabled the State Department several times requesting that the U.S. government take action similar to that of the British in protecting ships under its flag. In his reply Acheson informed Isbrandtsen representatives that their ships had gone into the area with full knowledge of the "hazardous situation." The company alone was to assume any risks. He also added that he thought the British were not convoying or escorting any vessels.[3] This statement was consistent with the president's views on the matter.

Earlier that day the State Department received word that a British warship had forced a Nationalist warship to retreat after it attempted to stop a British oil tanker and cargo ship from entering Shanghai waters. When the Nationalist ship threatened to open fire on the British steamers if they refused to stop, the British warship trained its guns on the Chinese vessel until it backed down. The British merchant ships were escorted up the river into the estuary for some distance.[4] On the next day an Isbrandtsen Company vessel arrived in the area, where it was ordered to halt by a Chinese naval ship. The American ship's master ignored the signal and continued up river within 100 yards of the Chinese ship. No further action was taken. A British warship had been following behind the American vessel, but it had done nothing.[5]

On the day following this incident Acting Secretary of State Webb related to the president the details of the British policy on escorting its vessels. Truman was informed of the official policy spelled out earlier by Acheson, but not of the incidents occurring earlier that week. Webb concluded his report by adding that the British government did not intend to break the blockade by force, although it had encouraged British merchant shipping to defy the blockade.[6] It remains a question why Webb did not relate the details of British action in China to Truman, but British intentions were clarified on November 1 when the Foreign Office notified the State Department that it was considering granting de jure recognition of the new authorities after consultations with other governments. The British explained their position:

> The United Kingdom has . . . to consider its own trading interests in China, which are considerable and of long standing. His Majesty's Government have advocated a policy of keeping a foot in the door, and if this policy is to bear fruit it can only be as a result of recognition of the Chinese Communist government. On political and practical grounds His Majesty's Government are therefore in favour of *de jure* recognition.[7]

This policy statement was an answer to pressure by British business interests in China, which had been urging recognition at the earliest possible date. Leonard L. Bacon, second secretary to the embassy at Nanjing, reported to Acheson on November 8 that "not only do virtually all responsible personnel [in the] British Embassy constantly and publicly state their expectation of recognition within one or two months, but some also fret at Foreign Office caution and shortsightedness which they imagine can be [the] only conceivable obstacle to it." Bacon also noted that the Chinese Communists have "subtly encouraged this attitude in definite preferential treatments." For example, labor troubles within British businesses had received less publicity and were settled with less fuss, and there were no problems comparable to the Ward case. But, according to Bacon, "when similar cases arise or are produced with Americans, they are protracted indefinitely and every drop [of] propaganda [is] squeezed out. This adds up to definite effort, not only to prevent rapid American recognition by producing [an] atmosphere in which it [is] impossible, but also to split [the] United States and Britain on this matter."[8]

The British held a conference in Singapore from November 2 to 4 to discuss the recognition question with other United Kingdom governments. The Bukit Serene Conference concluded that British interests in China and Hong Kong demanded the earliest possible de jure recognition of the PRC. The participants also concluded that recognition should be accompanied by a strengthening of resistance to the spread of communism in that area.[9]

Also at this time the American embassy in France reiterated that the French position on recognition followed the British. One French official had commented "that the safest way to stop the Communist armies at the Tonkin frontiers is to tie the fate of Indo-China to that of Singapore and Hong Kong. This could be accomplished by a simultaneous declaration of recognition by Great Britain, the Commonwealth, France, and the United States." The French Foreign Office furthermore had been informed from "various sources" that Mao Zedong was willing to offer certain guarantees in return for rapid recognition. It was believed that after recognition, the Chinese would not commit "the grave error" of invading Indochina until Mao's position in China had stabilized.[10]

With the governments of many European and Asian countries, such as France, Italy, Australia, India, and the Netherlands, responding favorably to the British policy, the State Department reinforced its efforts to assure that the Western powers would consult with each other before making any moves toward recognition and that it be agreed that the PRC should "respect at least the minimum standards of international conduct and to assume the responsibilities of a government in the treatment of foreign nationals and their interests." The British balked at this attempt, however, because their relations with the Nationalists had become even more strained. On November 4 the Nationalist Air Force announced to the British and American embassies in China that it had received instructions to bomb all shipping in the Formosa Straits. The British reacted by claiming that such a move was illegal and an unfriendly act. They countered by informing the Nationalists that as of November 5 their naval vessels and military aircraft would no longer be permitted to refuel or obtain supplies in Hong Kong, Kowloon, and the New Territories. When the State Department also lodged a protest against unprovoked bombing of American shipping, the Nationalists responded that there was little need for the Americans to worry; the focus of the move was the British, because of rumors circulating that Communist authorities had purchased British ships and were flying British flags.[11]

The CIA in Singapore also learned of an alleged Nationalist offer to the British. After two days of negotiations on October 22 and 23 between Nationalist Foreign Minister Ye Gongzhao and British Ambassador Ralph Stevenson, Ye had promised that the Nationalist government would make the following concessions in return for British nonrecognition of the Chinese Communist government: (1) relaxation of the Shanghai blockade so that British trade could be resumed; (2) reopening of the Yangzi River area to foreign shipping, if and when the Nationalists regained control of that area; (3) negotiation of a new Anglo-Chinese commercial treaty; and (4) a promise to make no claims on the British for the colony of Hong Kong. The Foreign Office had replied that it was unlikely that any offer would deter or delay recognition at this late date.[12]

Acheson and Truman Disagree

In light of the situation in Mukden, the problems between the British and the Nationalists, the British desire to accord de jure recognition to the Communists, and the unsettled situation concerning American business interests in China, the NSC undertook a review of its policy regarding trade with China at the request of President Truman. The study, completed on November 4, reviewed the recent developments in China, noting that the refusal to permit the withdrawal of the consulate staff from Mukden and Vice Consul Olive's arrest and beating by Chinese Communist police in Shanghai were "the most flagrant violations of international comity." On the other hand, it said that "no Americans have thus far been killed in Chinese Communist areas nor has any outright expropriation of American property been reported." The problem of imposing control over military-type exports was discussed. The United States had failed to employ such controls because of its inability to come to an agreement with the British on the issue and "because of the adverse effect it would have on United States business interests without compensating gains vis-à-vis the Chinese Communists."[13]

Among the report's recommendations was that the NSC policy toward trade spelled out in February should not be revised since it provided a sufficiently broad scope for a flexible policy. Acheson informed Admiral Souers that the NSC should be prepared for differences of opinion among government agencies concerning the continuation of this trade policy with the new government in China, especially since it was the president who had ordered a revision of the policy. Acheson recommended that several studies be undertaken to iron out any differences that might arise during a debate on the policy as outlined.[14]

Acheson was aware that a policy of continued trade with the Communists would not find favor with the president. Indeed, during the following week, Truman offered another plan to deal with the situation in Mukden. In a meeting with Under Secretary of State Webb on November 14, the president said that he felt that the U.S. government should thoroughly explore the possibility of blockading the movement of coal down the coast of China to Shanghai. He felt that if shipments of coal from Tianjin and other ports were prevented, "the Communists would understand that we meant business, and release Ward." He also indicated that he thought the move would gain considerable respect internationally and, futhermore, it would make it more difficult for the British to act independently in dealing with recognition. He said he was prepared to sink any vessels that refused to heed the American warnings.[15]

Acheson suggested to the president that a blockade of coal shipments would be futile since there was no evidence that Shanghai was receiving significant shipments from Tianjin or the northern port of Qinhuangdao. Shanghai received its coal from Anhui and Hebei, principally by rail. Moreover, the Nationalist government had been carrying out its blockade of central and southern ports since June, and as a result, coastal shipping was already at a low level. The remainder of such traffic was conducted with junks, with which it would be difficult, if not impossible, to interfere. Although the northern port of Tianjin was relatively active and not controlled by the Nationalist blockade, most of its traffic was with Hong Kong and under the authority of the British. Acheson noted that it would be illegal to stop British-owned ships, and stopping only Chinese ships would have a limited effect on conditions in China while supporting the Chinese argument concerning the imperialistic intentions of the United States.[16]

Acheson's reply to Truman and their obvious differences in opinion symbolized the dilemma faced by the State Department in dealing with the new Chinese authorities. On the one hand, the situation in Mukden was perceived as urgent, exasperating, embarrassing, and apparently directed only toward the United States. On the other hand, the remaining consulates in China had not been treated unfairly, and it was uncertain whether Beijing authorities supported or condoned the behavior in Mukden and other isolated cases. The hard line as suggested by Truman would not only threaten American interests in China but would contradict a policy followed throughout 1949 of allowing trade with the Communists in North China. At the same time that Truman wanted to halt coal shipments by force, the National Security Council suggested no change in the trade pattern carried on by American oil companies. For the time being, despite Truman's wish to use some form of direct action in getting the Americans out of Mukden, the policies that had been followed throughout 1949 were continued.

One reason for Truman's acquiescence on this issue was his meeting on November 17 with several consultants from the Office of Far Eastern Affairs,[17] after which he told Acheson that "he had gotten a new insight into the reasons for the Communist success in China, a better understanding of the whole situation, and found himself thinking about it in a quite new way." He added that he would like to meet again with the group and "set aside several hours for a discussion and go into all phases of the Far Eastern questions and policies." Acheson tried to explain the policy options to the president:

> Broadly speaking, there were two objectives: One might be to oppose the Communist regime, harass it, needle it, and if an opportunity appeared to attempt to overthrow it. Another objective of policy would be to attempt to detach it from subservience to Moscow and over a period of time encourage those vigorous influences which might modify it. [He] pointed out that this second alternative did not mean a policy of appeasement any more than it had in the case of Tito. If the Communists took action detrimental to the United States, it should be opposed with vigor, but the decision of many concrete questions would be much clarified by a decision as to whether we believed that we should and could overthrow the regime, or whether we believed that the second course outlined above was the wiser. [He] said that the Consultants were unanimous in their judgment that the second course was the preferable one.

The president concluded by stating that "in the broad sense in which [Acheson] was speaking that this was the correct analysis and that he wished to have a thorough understanding of all the facts in deciding the question. He believed that today's meeting had greatly helped him."[18]

But Acheson's optimism for this new attitude was short-lived. The contention between top officials in the U.S. government over the issue of trade with the Communists was summed up in a memorandum by Max W. Bishop, State Department representative on the NSC. In March 1950 the State Department gave permission to Standard-Vacuum and Caltex to ship 15,000 tons of motor gasoline, 5,000 tons of diesel oil, and 800 tons of lubricating oils to the PRC.[19] Some members of the government, in particular Secretary of Defense Forrestal and Secretary of Commerce Charles Sawyer, wanted this stopped. Bishop assessed the debate in this way:

> As I see it, [the] Defense position is essentially that the security position of the United States in Asia in particular and in the world in general requires that every effort be made to deny to Communist-China any economic assistance from non-Soviet sources, that accordingly all trade between the United States and Communist-China in strategic items . . . should be presumptively denied and should be allowed only in those instances where it can be shown on a case-by-case basis that no benefit would accrue to Communist-China. On the other hand, the Department of State position would seem to allow trade covering "normal civilian requirements" in non-strategic items . . . and to allow certain trade, on a case-by-case basis where it would be in our interest to do so or where the trade would not contribute to the military strength of the Communist block.

During the course of the debate, Acheson explained to Johnson and Sawyer:

> It is my view that Communist China should be treated as a satellite of the USSR and controls over exports to that area should be identical in scope and governed by the security principles now applied to the USSR and its eastern European satellites. However, the accomplishment of this objective clearly must, as a practical matter, be modified in those cases where the unilateral denial by the United States would merely surrender the Chinese market to the British and other suppliers.[20]

Thus Acheson was attempting to "keep a foot in the door," as the

British had done, despite protests by other top governmental officials.

U.S. Offers and Ultimatums

On January 5, 1950, President Truman announced that the United States was prepared to abandon Taiwan:

> The United States has no predatory design on Formosa or any other Chinese territory. The United States has no desire to obtain special rights or privileges or to establish military bases on Formosa at this time. Nor does it have any intention of utilizing its armed forces to interfere in the present situation. The United States Government will not pursue a course which will lead to involvement in the civil conflict in China.
>
> Similarly, the United States Government will not provide military aid or advice to Chinese forces on Formosa. In the view of the United States Government, the resources on Formosa are adequate to enable them to obtain the items which they might consider necessary for the defense of the Island. The United States Government proposes to continue under existing legislative authority the present ECA program of economic assistance.

Continued trade served the dual purpose of allowing some contact between the United States and the Communists, which was lacking in the diplomatic sphere, and giving the British competition in oil sales. Robert Aylward, vice consul in Beijing in 1950, has commented that Americans remaining in China perceived Truman's statement as an "olive branch" to the CCP.[21] The uproar it created in Washington would also demonstrate that many others agreed. The ultimate failure of this policy was aided by dissension within government circles, events in China during the first six months of 1950, and the State Department's attempts to have the "best of both worlds"—relations with the Communists and support for the Nationalists on Taiwan.

Twenty-four countries had recognized the PRC by mid-January; those that had not yet extended recognition were notified on

January 6, the day of the British recognition, that on January 13 all former military barracks areas of foreign governments in Beijing would be requisitioned.[22] In a policy statement prepared by the State Department on January 10 and approved by Truman that same day, it was recommended that an attempt be made to enlist British support in using a common approach to challenge the Chinese authorities' proclamation. But since this might be difficult, the State Department should request the British government to convey to the Chinese foreign minister that the United States was prepared to: (1) close all of its official establishments and withdraw all official personnel from Communist China if the requisition order was carried out; and (2) voluntarily return the property to the west of its consular compound (the "military barracks"), including immediate occupancy of the building there, and enter into negotiations regarding indemnification for the building. Clubb was notified in a secret telegram that if the British delayed taking action, he should seek out Zhou Enlai or other top Communist officials as a last resort and communicate this information to them. He was also to smuggle out or secretly destroy the consulate's confidential files and remove code materials to a safe place.[23]

Clubb assessed the motivation behind the move by the Communist authorities in Beijing. He felt that they did not want to create an international incident with the wholesale withdrawal of American personnel but were desirous of hastening recognition. He also assumed that this would bring the day closer when China and the Soviet Union would be seated together on the Security Council in the United Nations, thus breaking the USSR's isolation within this organization.[24]

The ultimatum presented by the State Department to the Communist authorities in Beijing was the fatal blow to the U.S. chances for an early recognition of the Communists. Because neither side backed down on this issue, it was impossible for an improvement of relations to take place. This was precisely what some Chinese wanted, given the continued U.S. support for the Nationalists on Taiwan. Robert Aylward has described the situation in the capital during January 1950: Mao Zedong had been in Moscow at the time the requisition order went out, and Beijing was being run by the CCP's Military Control Commission. According to Aylward, Beijing was "overrun" by Russian military advisers. Their presence was seen everywhere. Shops, particularly tailor shops, posted signs saying "Russian spoken here," and one could observe Russians daily buying Chinese goods. Aylward has speculated that it was the Russians who wanted foreigners, particularly Americans, out of Beijing. Within a week after Truman's January 5 statement, which was considered a peace offering to the CCP and a possible dissociation from Chiang Kai-shek, signs appeared on the walls of foreign legation quarters saying, "Military barracks of imperialist powers must be turned over to the PRC within seven days." Within a day or two, Aylward remembered, the British recognized the PRC and their sign disappeared. The Americans in Beijing waited for their property to be confiscated. On January 13 at 9 a.m., Communist authorities telephoned to notify the Americans that they must get out of the Old Main Compound, a former marine barracks then being used by the language school. During the day, the Americans moved the contents of the entire building, including seven tons of coal, to the building next door. During the late afternoon, Communist authorities telephoned to say that the Americans could have an extra day, but at 7 p.m. they changed their minds and said the Americans must be out by midnight. On the following day, the keys were handed over to Beijing authorities.[25]

As a result of this action, the State Department decided to close its embassy in Nanjing and other consulates throughout China. On January 14 it officially announced its intention to withdraw all U.S. personnel from China.[26] The Americans did not leave immediately. On the contrary, attempts to deal with the Communist authorities continued. Clubb requested an appointment for discussions with Zhou Enlai or another top official. Some of the Americans in Beijing trickled out of the city to return home during the following few months. At one point, CCP authorities tried to speed up the process by purchasing tickets for the Americans.²⁷ Trade between American oil companies and the China Petroleum Company continued. During the first week of January, for example, another Isbrandtsen Company vessel entered Shanghai, where it was shelled by the Nationalist navy in a lengthy and severe attack. The ship was rescued by a British sloop, but its cargo never reached its destination in Shanghai.²⁸

The Nationalists Threaten the American Position

The Nationalists began air raids on the Shanghai area in February. The Standard Oil installation in Shanghai and the American-owned Shanghai Power Company were partially destroyed in the bombings. Acheson ordered the embassy in China to register strong protests against these attacks not only on American companies but also on a large civilian population. One effect of the bombings was that the United States was blamed for what the Communists referred to as "imperialist murder." A February *Dagong bao* editorial depicted "American-made planes dropping American-made bombs, purposelessly incinerating the congested wooden neighborhoods" of Shanghai.²⁹

Sprouse, of the Office of Chinese Affairs, expressed his concern about the impact of these bombings on U.S.-China relations. In a memorandum to Merchant he concluded that "It seems incredible . . . that we permit the Chinese government brazenly to do the damage to our position in China that it is doing. It is all the more incredible that we do not take stronger steps than we have already taken to make clear to that Government that, if it continued bombing attacks of this nature, we will stop all aid or at least all shipments of military supplies from this country. We cannot afford to let the Chinese government take us further down the primrose path than it has already led us."³⁰ Robert C. Strong, chargé at Taibei, noted after one of his protests to the heads of the Nationalist navy and air force that both Chinese leaders had the idea that the United States would not retaliate in any serious way "almost regardless of what they do."[31]

But not all segments of the U.S. government criticized Nationalist attacks or the blockade of Shanghai. The CIA questioned the legality of alleged British aid to American shipping in Shanghai waters. The president was notified in January that in the American view, the Nationalist navy and air force were agencies of a friendly government so long as recognition was extended. Any action against those agencies by either American or British ships should not be condoned.

The CIA also predicted more problems for Anglo-American relations if the British continued to act against the Nationalist navy. The issue was brought out at a meeting of British Ambassador Oliver Franks, Acheson, and other State Department members in March. The results of this meeting not only signaled the deep disagreement between the two governments on this issue, but also dashed any hopes of formulating an American-European front concerning controls over strategic materials trade with China. Franks sought the answers to two questions: Would the United States use its influence to terminate the Nationalist blockade and bombings? And what would the State Department's attitude be if the British themselves undertook to break the blockade? The British Foreign Office believed that it was in the interests of both Britain and the United States to stay in China as long as possible and to maintain contacts with the Chinese.

In reply Acheson pointed out that British actions in China were jeopardizing Anglo-American relations. The governor of Hong Kong, for example, had allowed the Chinese Communists access to spare parts and machinery while denying similar goods to the Nationalists. Eight hundred tons of this equipment had been transported on a British ship to a Communist port. Furthermore, "it was clear that this was only a beginning and that this subject would become even hotter." Several congressmen had begun to look into this matter and were talking of curtailing postwar aid to Britain unless the problem was resolved.

The British ambassador replied that he viewed the matter differently. "In the first place," he stated, "from the cables he had seen, it seemed clear that the legal handling of the matter (the blockade) on the United States side had been very inept." He was also aware of the "heat" that had been generating in Congress, and he wished to point out that Bevin had a similar problem with Parliament because the planes that were being used to destroy British lives and property in Shanghai had been supplied by the United States. No conclusions were reached at this meeting, and both Franks and Acheson said that they would avoid discussion of the issues with the press.[32]

Disagreements with the British were not the only problems for Acheson and others who thought that an accord with the Communists might be possible. On February 14 the government of the Soviet Union and the PRC signed a thirty-year Treaty of Friendship, Alliance, and Mutual Aid. Despite State Department assessments that negotiations between Mao and Stalin since November 1949 had been long and difficult, with the Chinese holding out for the best possible deal, the fact that the treaty was ratified and put into force in April seemed to confirm American suspicions that the Kremlin was in charge in Beijing. According to the Office of Far Eastern Affairs, it was believed that the "critical objective" of the USSR in East Asia was to consolidate and perpetuate its control over China.[33] The CIA's assessment of the treaty was that it would not be "so transparently exploitative in character" that Mao could not justify it to the CCP or the Chinese people, but it would strengthen the Soviet position in China. "The USSR, exploiting China's need for 'friendly aid,' can gradually strengthen the Soviet hold on the party machine, the armed forces, the secret police, communications, and channels of information. . . . On balance it would appear that the USSR, mindful of previous failure in China and conscious of the importance of a Chinese Communist ally as a bulwark against U.S. power in the Pacific, will continue, both cautiously and with determination, its planned infiltration in Chinese political, military, and economic affairs."[34]

The American Embassy Is Closed

Perhaps the one event that made the State Department finally decide that it was time for Americans to leave China was the meeting between Clubb and an official named Lin from the Alien Affairs Office in Beijing. The months of requests for an interview with Zhou Enlai or a top official in Beijing were finally honored in April 1950 after the British undertook efforts on Clubb's behalf. Clubb discussed the general political, economic, and social problems encountered by Americans in China, citing specific examples of mistreatment of some Americans. Lin replied that his time was short and characterized what Clubb said as "worthless talk." He also stated that the cases Clubb mentioned as violations of China's international obligations were actually U.S. failure to respect China's sovereignty. He concluded that "so long as [the] United States continued to support Chiang Kai-shek, talk of working an improvement in [the] general situation was ridiculous." Clubb concluded from his conversations with Lin that U.S. problems in China would be cleared up when the issue of support for Chiang was resolved; otherwise, they would not be solved.[35]

On April 10 the American consulate in Beijing was closed. Clubb turned over custody of the building and moveable government property to the British. Codes, seals, and confidential materials were destroyed. It was not until June that increased trade restrictions were placed on goods in China. In a letter to the Secretary of Commerce Sawyer, Acheson noted that "the Department of State now believed it necessary to apply export licensing controls over strategic materials for shipment to Communist China on a more restrictive basis than hitherto." There would be two exceptions, however. Licenses would still be granted to American exporters in cases where their denial would merely divert the trade to suppliers in Britain or Western Europe, and where these European suppliers were already trading substantial quantities of certain materials.

With this strengthening of the old policy toward trade with

China, Acheson was attempting to impose unilateral controls over strategic exports while still allowing Caltex and Standard Oil, for example, to compete with European companies, particularly British-owned Shell Oil Company. But the outbreak of the Korean War on June 25 and the subsequent UN Security Council resolution to impose an embargo on all exports to North Korea undermined this revised trade policy. On June 29 Caltex and Standard oil companies were asked to suspend all shipments of petroleum products to China for the time being.[36]

While the American companies complied with the request, according to Commerce Department sources, a similar petition to the British was rebuffed. Anglo-American relations continued to deteriorate: Acheson commented on July 12 that he was "astonished" that the United Kingdom could not agree to apply controls over trade with Communist China considering the support in the Chinese media for the North Korean "aggressors." It appeared, moreover, that the British were developing a policy clearly contrary to that of the United States. During the first week of July, the Foreign Office had specifically instructed Shell Oil Company to continue "normal sales" to South China ports but not to make bulk shipments to North China, and on July 11 Shell was informed that it could resume bulk shipments even to North China. During the week of July 12, 700 tons of motor gasoline were sent to Tianjin, and a second tanker left for Qingdao. The State Department noted that "both these ports are, of course, in North China and very close to Korea."[37]

The Foreign Office gave three reasons for its reluctance to stop all oil shipments to China. It feared that such action would provoke retaliatory moves against Hong Kong, where the oil companies' headquarters were based. It might further prejudice relations with the new Chinese government and thus jeopardize British economic interests in China. Furthermore, the Foreign Office claimed that Shell's shipments were an "insignificant trickle" and based on civilian requirements, especially since Shell officials had promised, under State Department pressure, not to accept any contracts that would have gone to American companies.

There was yet another problem. On July 21 Shell officials in Hong Kong accused the American companies of continuing trade with the Communists through Hong Kong brokers. British sources claimed that gasoline was being shipped from the United States via Mexico and South America to North China ports. Although a State Department investigation revealed no evidence of such transactions, Acheson agreed to allow the British Admiralty in Hong Kong to requisition all American-owned oil stocks coming through the colony in order to control their reexport. This meant that the American companies as well as Shell could not distribute their stocks without the approval of the Admiralty. Both Caltex and Standard Oil officials reacted favorably to this set-up because they preferred to be treated the same way as Shell so that they could argue to the Chinese Communists that they had no choice but to cut back trade. This also implied that the Admiralty might allow reexport to the Chinese. The Foreign Office indicated that if it became evident that there were stocks in excess of Hong Kong's needs, it would be difficult to defend stoppage of shipments to China.[38]

This policy resulted in a public outcry in the United States for the State Department to do something to stop the British from shipping oil to China. In a petition sent to the White House in the summer of 1950, seventy-one members of Congress stated that "we have noted with a great deal of concern that while we have acted to resist the aggression in the case of North Korea . . . we still continue to trade and 'do business' with the 'seconds' or supporters of these same aggressors. Why?" The petition demanded strong action against any country that traded with supporters of the North Koreans. Nonetheless, nothing further was done until after the Chinese entered the Korean War in November. The Department of Commerce was then forced to react to yet another congressional protest with a denial that shipments of strategic commodities had been permitted to go to China after the outbreak of war in June. On December 3 strict controls were placed over shipments of foreign origin that entered U.S. ports or passed through any areas under American jurisdiction en route to the USSR, its satellite countries in Europe, China, Hong Kong, and Macao. The directive specifically mentioned that Hong Kong and Macao were included because of their importance as transshipment points. Any such trade must be licensed by the Commerce Department's Office of International Trade. In this way, that office could prevent any American trade with China as well as maintain a more careful watch over British and other foreign shipments that passed through the Panama Canal or stopped at islands in the Pacific under American jurisdiction.[39]

Conclusion

The outbreak of war in Korea has been considered the final issue that prevented U.S. recognition of the People's Republic of China in 1950 because of the association of Communist agression on the Korean peninsula with the Soviet Union and the Chinese Communists. But there remain some problems with this conclusion. Sino-American relations were troubled in the years leading up to the outbreak of the war. The two biggest issues of contention were the continued U.S. association with the Nationalist regime even after its defeat on the mainland and retreat to Taiwan and the Communists' apparent mistreatment of the American consulate staff in Mukden and other isolated incidents. Because of American economic and military aid to Chiang Kai-shek's group, the Communists perceived that American intentions were aimed at their defeat and later, when victory was secured, at prolonging war and saving the remnants of the enemy regime in an isolated province. When the overcrowded, overwhelmingly poverty-stricken population of Shanghai was bombed with American planes and bombs during early 1950, the United States government lost whatever meager credibility it had had in the eyes of Chinese leaders.[40] These actions as well as the delay in according either de facto or de jure recognition laid to rest any hope of securing the support of that group of CCP leaders who wanted American aid to counterbalance Soviet influence, to whom Zhou

Enlai had referred in his secret message.

From the American point of view, the delay was necessary because of the situation in Mukden, which made it impossible for the State Department to consider recognition. The Truman administration balked at giving into Communist pressure, which the British had done as early as March 1949. As the problem worsened, with the repeated failure to meet with CCP leaders, the arrest of Ward, and the confiscation of some American property, it only became harder to work toward normalized relations. Acheson had stated that it was impossible to deal with the question of recognition because of these issues.[41] Moreover, in early 1950 the Truman administration was under attack from Nationalist supporters in Congress, the press, and the American public. Admiral Sidney Souers, head of the NSC, recalls that the China question was emphasized by anti-administration Republicans after Truman's 1948 "surprise" electoral victory and especially in conjunction with the 1950 congressional elections.[42] This has been considered an important ingredient in U.S.-China relations. Because of the domestic pressure, Truman hesitated to terminate aid to the Nationalists and prevented steps that might have led to recognition.[43]

Even if the problems in Mukden had not occurred, it is still unclear whether the United States would have been successful in normalizing relations with the People's Republic of China in 1949 or 1950. The State Department's approval of continued trade with the Communists and the maintenance of the American-run Shanghai Power Company until after the outbreak of the Korean War, accompanied by Acheson's attempts to ensure American economic competition with the British in China, point to the idea that efforts would have been made to recognize the Communist government for the protection of American economic interests. On the other hand, it is also unlikely that the United States would have abandoned Chiang Kai-shek. A policy of using Taiwan as an anti-Communist bastion predated even Chiang's move to the island. American objectives in East Asia included promoting freedom of trade on the mainland and maintaining an "island security belt" off the coast of East Asia.[44] But these two objectives were not seen as mutually exclusive. In January 1950, the CCP had accepted de jure recognition by the British while allowing them to maintain a consulate on Taiwan. The State Department had publicly divorced itself from the island's fate with Truman's January 5, 1950, speech. Plans were formulated within the State Department to keep Taiwan from the Communists through the United Nations, without unilateral action by the United States, an option similar to that carried out in Korea in June 1950. Whether the CCP would have believed the public statements and accepted an understanding similar to the British solution given the history of American aid to Chiang can only be pondered.

Several writers have debated this question and have concluded that Chinese-American hostility and the eventual break were inevitable even before the outbreak of the Korean War because of the apparent affinity between CCP leaders and Moscow and the anti-American actions perpetrated by the Communists throughout China.[45] Nonetheless, both sides had a similar motive for maintaining relations: for the CCP, the United States could have played a role in the improvement of the war-torn Chinese economy; for the United States, recognition of the new government would have allowed American economic interests to protect their investments in China. American personnel, for example, did not abandon the Shanghai Power Company until December 28, 1950, when the Communists took over and named their own managers to the plant.[46] This was more than a month after the Chinese intervened in the Korean War and the U.S. government had publicly condemned them as aggressors.

Diplomatic relations, on the other hand, continued to deteriorate beginning in late 1948 with the holding of Ward and his staff in Mukden. The Truman administration would not give in to such pressures, and the Chinese Communists refused to acquiesce without the promise of at least de facto recognition of their authority in North China, which they had received from other Western powers as early as March 1949. As time passed, relations only worsened while attempts to work out some sort of accommodation continued. The State Department tried to contact CCP leaders to discuss the Mukden situation, Ward's captivity was kept from the American press for fourteen months until a few weeks before his departure to prevent adverse publicity,[47] and the consulates in China were not abandoned even after several incidents involving American citizens. The Chinese approached the United States through the Zhou demarche and, perhaps more significantly, Huang's invitation to Ambassador Stuart. Although the American consular communities were not accorded diplomatic status because of the refusal to recognize the Communists, few American "private citizens" in China were victims of Communist hostility.[48]

But neither side backed down from its demands. How long would the U.S. government tolerate the holding of American diplomatic personnel as prisoners? How long would the Chinese Communists wait to be recognized by the United States after other Western nations had already done so? These are the crucial questions for Sino-American relations in 1949, but they do not point to an inevitable split between the two countries.

O. Edmund Clubb, consul general in Peking in 1949, also finds it difficult to agree with the premise that a Sino-American breach was preordained in 1949. He asks the reader to "just suppose," for example, that Stuart's trip to Beijing in June 1949 was approved by the Truman administration or that diplomatic recognition of the PRC had been granted before December 1949 when Mao made his visit to Moscow.[49] Whatever the answer, it can be concluded that this chapter in U.S.-China relations colored developments for nearly thirty years between the two countries. It also influenced relations between the United States and Britain, each country following a separate path after 1949 when the explicit goal had been the formation of a common policy in East Asia to show that the non-Communist world was united against what was perceived as the spread of Soviet influence. The British, as well as other Europeans and Asians, dealt with the reality of the situation in China and, in the end, preserved their economic interests there.

7. Taiwan "Abandoned," Taiwan Preserved

In the first six months of 1950, until the outbreak of the Korean War on June 25, the Truman administration made preparations to abandon the Nationalists on Taiwan. This idea was clearly manifested in Truman's statement of January 5, in which he announced that the United States had "no predatory designs on Formosa." It was also at this time that the State Department sent orders to reduce the Taibei embassy to a skeleton staff.

Nationalist supporters in Congress denounced such actions and took every opportunity to challenge Truman's decision. On January 5 Senator William F. Knowland (R., California) opened the attack on the president on the Senate floor by calling for a major shake-up in the Far Eastern Division of the State Department, the appointment of General Douglas MacArthur as coordinator of American policy in East Asia, and the dispatch of a military mission to supervise aid to Taiwan. Senator Robert A. Taft (R., Ohio) claimed that to reject the idea of using American military forces to stop the spread of communism to Taiwan was inconsistent with what the United States was already doing in Europe. Senator H. Alexander Smith (R., New Jersey) proposed an American occupation of the island, by which, he later explained, he meant the establishment of a joint political authority on Taiwan. Senator Arthur S. Vandenberg (R., Michigan), who had earlier worked with the administration on policy toward

China, issued a statement to the press expressing his regret that appropriate committees in Congress had not been consulted. But these statements apparently represented a minority opinion since Acheson, at a January 6 cabinet meeting, commented that he thought the "reception of [the] statement of Formosa ha[d] been favorable."[1]

Within a week, on January 12, Acheson addressed this issue again in a speech before the National Press Club. He outlined an American "sphere of strategic concern," a "defensive perimeter," which included Japan, the Philippines, and surrounding islands, but did not mention Taiwan or South Korea. Most Americans, particularly members of the China bloc, interpreted this as a signal of the "abandonment" of these two excluded areas.[2]

During the next few months some members of the State Department began to push for a change in this policy. Assistant Secretary of State for Far Eastern Affairs Dean Rusk sent a memorandum to Acheson on May 30 suggesting that if the United States did not act and announce that it would "neutralize" Taiwan, it would symbolize another retreat in the face of Soviet aggression. Rusk concluded that "Admittedly, a strong stand at Formosa would involve a slightly increased risk of early war. But sometimes such a risk has to be taken in order to preserve peace in the world and to keep the national prestige required if we are to play our indispensable part in sustaining a free world. Action to be effective must be prompt."[3]

John Foster Dulles, consultant to Acheson, sent an identical memorandum. Earlier Dulles had suggested that the United States should take over Taiwan and "make a show-piece" of it.[4] He also sent other memoranda during his visit to Japan in June 1950 relating MacArthur's belief that Taiwan was not only strategically significant to the United States but was also important as "a political area to western ideology." MacArthur had asserted, "I am satisfied . . . that the domination of Formosa by an unfriendly power would be a disaster of utmost importance to the U.S., and I am convinced that time is of the essence."[5]

The proponents of action were satisfied on June 27, following the outbreak of fighting in Korea, when the president issued a statement saying:

> I have ordered the Seventh Fleet to prevent any attack on Formosa. As a corollary of this action, I am calling on the Chinese Government on Formosa to cease all air and sea operations against the mainland. The Seventh Fleet will see that this is done. The determination of the future status of Formosa must await the restoration of security in the Pacific, a peace settlement with Japan, or consideration by the United Nations.

Historians have also viewed this statement as the reflection of a fundamental change in American policy toward Taiwan.[6] A case can be made, however, for the argument that the Truman administration did not abandon the Nationalists on Taiwan in January but operated under the policy formulated by mid-1949 of preserving that island against the Communist advance. But this policy was kept secret from lesser State Department officials, congressional leaders, the American public, and the Nationalists on Taiwan for the same reason outlined earlier for the release of the White Paper—to push the Nationalists into a strong, independent posture and to keep the lines of communication open with the new Communist administration. Those who knew of this policy were the officials whose efforts had been coordinated by Acheson beginning in February 1949. Included among this group were the president and the secretary of state himself; Acting Secretary of State Robert A. Lovett; the Joint Chiefs of Staff; the head of the National Security Council, Sidney W. Souers; W. Walton Butterworth of the Office of Far Eastern Affairs; and "Special Representative to Taiwan" Livingston T. Merchant.

It is also apparent that MacArthur, Rusk, Dulles, and Secretary of Defense Louis Johnson were not aware of any such policy. Johnson was an outspoken critic of the State Department's Taiwan policy during this time.[7] Moreover, Edwin W. Martin, consul at Taibei from October 1949 to January 1950, has stated that the United States "had no program . . . no policy to try to keep Taiwan out of Communist hands. This was not the policy. Our function there was strictly observing and seeing what was going on." Martin asserted that the president's statement in January seemed to reflect that the Communist victory was accepted "with a certain amount of equanimity." Taiwan would not be defended, which was why he left the island. "It was anticipated that the Communists as early as March of '50 might launch an invasion."[8]

Martin also stated that the United States "had no military aid program then of any kind there. We had long ago recognized that Formosa was part of China and that was it." Martin was apparently unaware of the many projects under way on the island both before and during his five-month stay there. When questioned about the State Department's policy of briefing a new administrator sent overseas he stated, "Well, in those days we didn't get much of this sort of thing like briefing and so forth. . . . You just were thrown in and told to swim." Martin was home on leave when Merchant was stationed in Taiwan.

Critical Allies

It is most interesting that, although Martin claimed that there was no American military aid program in Taiwan while he was there, both the British and Philippine embassies complained to the State Department about the continuation of shipments of arms and military supplies. On December 6, 1949, the British embassy sent a memorandum to the State Department stating that British sources revealed that "substantial quantities of American military equipment," including 100 tanks and 108 B-25 bombers, had arrived in Taiwan. It asked why this was done if the U.S. government "had concluded that no practical steps could be taken to prevent Formosa [from] falling into Communist hands." The memorandum concuded that "His Majesty's Government feel much concern at the above circumstances and hope that the United States Government will feel able to take steps to stop or restrict the flow of arms from the United States to Formosa."[9]

The reason for such concern had been spelled out at a meeting the day before between Counselor to the British Embassy H. A. Graves, Livingston Merchant, and Philip D. Sprouse. The British feared that much of this equipment would eventually fall into Communist hands, as did American equipment given to the Nationalists on the mainland, and then it would be used against Hong Kong. Graves stated that the French had also expressed this fear concerning the fate of Southeast Asia.[10]

A press despatch from Manila dated December 12 stated that the U.S. government had informed the Philippine government that "it will take positive action to counteract communist threat to Formosa" and that the United States would send Chiang adequate arms to enable him to hold the island. Acheson reacted to this information by asking Myron M. Cowen, ambassador to the Philippines, if he had "any information as to [the] source [of] this unfounded assertion." He further suggested that Cowen use his discretion in informing the Philippine government that the State Department found "such irresponsible statements difficult [to] understand and often embarrassing."[11]

On December 21 Cowen reported that the source of the story was President Elpidio Quirino, who had told the press that he believed that the Nationalists would not have to leave Taiwan because the United States "had lately shipped large quantities [of] military material to Formosa." He had not said, however, that the United States would take "positive action" in the area. Several American correspondents had also questioned Cowen after their interview with Quirino. He informed them that he knew nothing about the "alleged recent U.S. shipments [of] arms aid to Taiwan." But he immediately tried to ascertain where Quirino got his information and came to the tentative conclusion that the source was either Dorothy Brandon, a correspondent from the *New York Herald Tribune* who had allegedly passed through a Taiwanese port recently, or the Chinese ambassador to the Philippines. Cowen concluded that "President Quirino is prone to sound off without checking his facts and to state as fact that which he wished to believe, and I am inclined [to] think his credulity and wishful thinking also were responsible for [the] character of [the] statement made during interview in question."[12]

Secret Aid Continues

While the State Department was denying the existence of continued military assistance to Taiwan, Butterworth prepared a memorandum for Acheson listing the military supplies that remained to be shipped to Taiwan. Some of the supplies were paid for by the $125 million grant; other items were bought by the Nationalists with their own funds. Tables 2 to 4 list the military equipment sent to the Nationalists after November 1, 1949.

Assessments of Nationalist Strength

On December 7 the Nationalist government announced the appointment of K. C. Wu, former mayor of Shanghai and close associate of Chiang, as the new governor of Taiwan to replace Chen Cheng, who planned to return to a military command position.[13] The Chongqing faction of the Nationalist government began arriving on December 8, and during the following weeks there was considerable movement within the Nationalist administration. Acting President Li was in New York and Chiang had resumed the presidency. On December 19 Acheson requested the consul at Taibei, Donald D. Edgar, to make an estimate of the following factors because "intelligence agencies [were] reviewing prospects [in] Taiwan":

    1. Present popular support [for the] Nationalist Government, Commies, Taiwanese independence leaders, and factors influencing changes [in] such support.
    2. Internal unity [of the] Nationalist Government including prospects [of a] coup by anti-Communist Nat[ionalist] Government milit[ary] cmdrs.
    3. Views [of] Nationalist leaders on survival prospects [of]

Govt and estimate [on] probability [of] defections to Commies.

4. Effect on General situation [of] (a) establishment [of] Nat[ional] Govt [on] Taiwan (b) appointment [of] K. C. Wu [as] governor.

5. Prospects [of] economic stability.

6. Official and public receptivity [of] (a) U.S. milit[ary] occupation (b) UN trusteeship.[14]

Table 2

Military Material Procured for Chinese Government under $125 Million Grant but Not Yet Shipped, as of November 1, 1949

	Quantity	Value
Light tanks	100 ea.	
Scout cars, M3A1	100 ea.	
Motor carriages, 75 mm. howitzer	125 ea.	
Rifles, auto., brng., M1918A2 (BAR)	1,000 ea.	
Mounts, combination, M23A1 for 37 mm. gun (used on light armored car)	100 ea.	
Shells for 4.2" chemical mortar	25,000 ea.	
Tires (various)	7,000 ea.	
Automotive spare parts		$1,680,000
Weapons spare parts		85,000
Tools and tool sets		240,000
Raw materials (inclusive of gun powder)		1,850,000
Antimony sulphide (for vulcanizing rubber)	5,000 lbs.	
Gyro stabilizers	100 ea.	
Dry batteries	40,000 ea.	
Radio sets	600 ea.	
Radio spare parts		6,000
Electrical equipment (including one 5 KW diesel-driven generator per set)	12 sets	
Medical supplies and equipment		995,000
Vacuum tubes		47,000
Naval spare parts (hull and engineering spares)		265,000
Aircraft spare parts		305,000
Plastic film		80,000

Source: Taken from Butterworth to Acheson, 7 December 1949, *FRUS*, 1949, 9:438–40.

Table 3
Estimated Dollar Value of Unshipped Items, by Federal Agency

Agency	Value
Army	$6,793,500
Navy	310,000
Air Force	390,700
Treasury—Bureau of Federal Supply	500.000
Total	$7,994,200

Source: Taken from Butterworth to Acheson, 7 December 1949, *FRUS*, 1949, 9:440.

Table 4
Military Material Procured by Chinese through Commercial Sources on Which as Yet Unused Export Licenses Have Been Issued, as of November 1949

	Quantity	Value
A. Items Other Than Aircraft and Aircraft Parts		
1. Items approved for export		
Drop steel forgings		$ 4,550
Tank spares	400 tons	60,000
Motor carriages tracks	40 sets	19,000
Motor carriages (in transit from U.K.)	30 ea.	75,000
Sherman tanks (in transit from U.K.)	85 ea.	300,000
Powder and ammunition		205,475
Propellant flakes and powder		668,520
Cartridges (in transit from Canada)	81,000,000 rnds	4,900,000
Ammunition, .30 caliber		21,281
Shot firing cord		4,570
Motor carriages	42 ea.	100,000
Light tanks and dozers	9 ea.	72,000
Mauser rifles & ammunition, 7.9 mm (in transit from Belgium)		4,215,000
Mauser rifles & machine guns, 7.9 mm (in transit from Belgium)		559,100
Telescopes and periscopes		2,300
Gun-sights		1,000
Gun-sight noise filters		45
Smokeless powder		3,355
Rocket igniters		5,500
Tank spare parts	200 tons	30,000

Table 4 continued

2. Items not yet approved for export but requested by Chinese Government

Used light armored cars	200 ea.	50,000
Used tanks with flame throwers	4 ea.	8,000
Tracked landing vehicles	156 ea.	75,000
Staghound armored cars (in transit from U.K.)	200 ea.	200,000
Rifles and machine guns (in transit from Belgium)		84,000
Machine guns (in transit from Belgium)	3,000 ea.	1,395,000

B. Aircraft and Aircraft Parts

There are still valid export licenses for approximately $6,800,00 worth of aircraft and aircraft parts purchased commercially by the Chinese Government. As this total includes some $2,700,000 for 180 A T6 aircraft which have been reported by the Chinese to have been shipped on a continuing basis throughout this year, it is believed that a good portion of the $6,800,00 total has already been shipped, though customs reports on such shipments have not yet reached the Department.

Source: Taken from Butterworth to Acheson, 7 December 1949, *FRUS*, 1949, 9:440.

Note: Some of the items listed are being procured with funds from the $125 Million Grants, but the exact quantity of such items is not known.

Edgar's reply was received on December 23. He concluded that the Nationalist government had no popular support and that its power rested in Chiang, the military, and the secret police. Communist support was limited to small groups with a flimsy organization. Taiwanese independence leaders were weak. There was little internal unity among the Nationalists, but Chiang turned this to his advantage. No general was permitted to acquire the strength needed to insure a coup. The official position within the Nationalist government was that it needed outside aid to survive—specifically Chiang wanted the $75 million allocated by Congress in September plus any unused ECA funds. The Taiwanese had reacted unfavorably to the establishment of the Nationalist government on Taiwan, and the appointment of Wu had resulted in "political confusion." The local economy was being wrecked by "unbridled military demands," but the island's economy was basically sound and possessed the necessary elements to make it self-sustaining once military demands were related to the ability of the economy to pay for them. As for the U.S. military occupation, the top Nationalist leaders would be expected to require a "face-saving formula" such as support for a return to the mainland, but American military control advisers were "entirely feasible" at this point since this had been suggested by Chinese officials. "Many [of the] highest Chinese officials have repeatedly requested maximum U.S. military cooperation, even supporting condominium." The mass of Taiwanese hated the Nationalists, feared the Communists, and "look[ed] hopefully" to a temporary American takeover. The idea of a United Nations trusteeship, moreover, was preferred by the Taiwanese but not by the Nationalists, who would view this as a loss of their power. But, Edgar added, "Chinese officials would accept any proposed formula."[15]

Ambassador Koo came forth with another request for aid on December 23, 1949. Koo's estimate of the situation on Taiwan differed considerably from that of Edgar. He related the kinds of land and tax reforms being carried out by Wu to assure the support of the local population. His request for aid was for the purpose of carrying a war to the mainland because the Nationalists' resources on the island "were sufficient for its own support and defense."[16]

On the same day that the State Department received Ambassador Koo's request, a memorandum entitled "Possible United States Action Toward Taiwan Not Involving Major Military Forces," prepared by the Joint Chiefs of Staff, was sent to the National Security Council for circulation before the proposed NSC meeting of December 29 to update U.S. policy toward Taiwan. Butterworth commented that the memorandum "parallels with extraordinary fidelity the request for increased assistance from the Chinese National Government received on the same date."[17]

In the memorandum, General Omar Bradley, chair of the Joint Chiefs, reported that an assessment had been made of military measures, short of the despatch of a major military force, that might be undertaken with respect to Taiwan "in furtherance of United States political, economic, and psychological measures now under way." The conclusions were reached that "a modest, closely supervised program of military aid" would be in the security interests of the United States, and that such a program should be integrated with a stepped up political, economic, and psychological program. If the National Security Council concurred, the Joint Chiefs also planned to direct the commander in chief for the Far East and the commander of the 7th Task Fleet to make an immediate survey of the extent of military assistance required to hold Taiwan in case of attack. After receiving this survey, the Joint Chiefs would make recommendations concerning a military program.[18]

At the December 29 NSC meeting Acheson challenged the conclusions in the memorandum, stating that it had been his understanding from past pronouncements by the Joint Chiefs that the "strategic importance of Formosa was insufficient to warrant the use of United States armed forces." This report seemed to give a different view on the matter. It was concluded during the discussions that followed that there was no intention to send combat troops to Taiwan, and it would be left to the State Department to determine the form any military aid would take.[19]

On the next day, Acheson sent a telegram to Kenneth C. Krentz, a member of the Policy Planning Staff then in Taiwan, for Krentz's "secret" information. It stated that the National Security Council meeting had reconfirmed "existing top policy re Formosa"—that is, that the U.S. government would continue to pursue political and economic means to deny Taiwan to the Communists because the Nationalists' resources were believed to be sufficient to accomplish this and an expanded American role there might be costly to the achievement of American objectives on the mainland. Krentz was instructed to cable any preliminary comments on the "general situation" "eyes only Butterworth."[20]

By late 1949 the State Department realized that most of its allies would be recognizing the Communist government, and it also was pursuing the possibility of this course of action. On the other hand, the hope for saving Taiwan from the Communists lay in Nationalist defensive strength. Shipments of arms, despite protests from the British and the French, continued.[21] Furthermore, the State Department had not completely ruled out the use of military force. The debate involving the preparation of Truman's January 5 statement reflected this.

There were two important changes made in Truman's statement the day it was released. Both changes were in a sentence that originally read, "The United States has no desire to obtain special rights or privileges or to establish military bases on Formosa or to detach Formosa from China," and both were suggested by General Omar Bradley of the Joint Chiefs of Staff, whom the president had specifically asked to read the statement before its release. Bradley requested that the phrase "or to detach Formosa from China" be deleted because "the situation might arise where they [the Communists] will march South, in which case we may want to detach Formosa from China." Acheson agreed to delete the phrase although he would have preferred to leave it.[22] The second revision was the addition of the phrase "at this time" to the end of the sentence, thus having it read, "The United States has no desire to obtain special rights or privileges or to establish military bases on Formosa at this time." Bradley wanted this change "because of the possibility, in the event of war, we might have to recapture bases on Formosa."[23]

From January to June 1950 the State Department pursued a policy of pushing the Nationalists toward greater independence from the United States while continuing the existing military and economic aid programs. This apparently confusing policy was not contradictory, as Acheson explained to Secretary of Defense Louis Johnson. Johnson had suggested that aid programs such as the China Aid Act were superseded by the president's January 5 statement and therefore must be discontinued. Acheson explained that the phrase stating that the United States would "not provide military aid or advice to Chinese forces in Formosa" was intended to prevent further supply of military supplies beyond the $125 million grant, which as yet had not been completely expended.

On the other hand, the phrase "resources on Formosa are adequate to enable them to obtain items which they might consider necessary for the defense of the island" was intended to allow the Nationalists to purchase material with their own funds.[24]

As a result, ECA aid and military aid from the $125 million grant continued.[25] On January 19, 1950, the ECA presented a twelve-page report entitled "Summary Statement and Program Requirements for the Period February 16, 1950-June 30, 1951" to the State Department. The purpose of this paper was "to set forth a program for the continuation of economic assistance for Formosa for the balance of the current [1950] fiscal year and for the fiscal year 1951." In a detailed analysis, the costs envisioned for the continuation of the ECA's four major categories of expenditure for its China program—a commodity program, industrial replacement and reconstruction projects, the Joint Commission on Rural Reconstruction, and administrative expenses—were outlined. Because the ECA's projects on the mainland had "undergone progressive contraction" from early 1949, when Communist forces surrounded northern Chinese cities, until November 1949, when the southern provinces of Sichuan and Guangxi were occupied, much of the money originally allocated to mainland cities had become available for Taiwan.[26]

As of February 15, 1950, $104 million of ECA funds had not yet been spent in China. Of this total, it was proposed that $26,170,000 be made available for continued use in the commodity program on Taiwan. This would subsidize shipments of fertilizer, cotton, and petroleum and provide for the sale of yarn manufactured from ECA-supplied raw cotton in Shanghai and taken from that city before the Communist occupation in May 1949. The funds from yarn sales as well as sugar trade with Japan would yield approximately $5 million, which was to be used to purchase other essential commodities, such as wheat, soybeans, fats and oils, and gypsum. It was also planned to use $480,000 for technical services for the program of industrial reconstruction and rehabilitation. This program's goal for fiscal year 1951 was to provide materials for transportation facilities on the island, such as railway cross ties, rails, and highway bridge steel. An additional $7,062,000 was to be reserved for assistance to the island's electrical power system, chemical fertilizer, and other industries. It was hoped that the Joint Commission on Rural Reconstruction would meet its operational expenses from local currency funds developed or traded, but it was planned to allocate an additional $70,000 for purchasing supplies and equipment needed until June 30, 1951. The commission had begun to devote "its full energies" to the program on Taiwan after it was forced to withdraw from the mainland after October 1, 1949. Administrative costs—in both Washington and Taibei—were to be allotted $550,000.[27]

In May the Joint Chiefs of Staff reiterated their position favoring the possible use of American military aid to Taiwan. In a report to the secretary of defense, they concluded that "successful resistance on the part of the Chinese Nationalists, particularly in the Formosa area, is in the military interest of the United States."[28] Furthermore, in June a request came from the Nationalists to purchase napalm bombs and other incendiaries with money from the $125 million grant. The Defense Department had initially refused, not because of the nature of the materials requested, but because there was some confusion as to the availability of funds. After this problem was resolved, the request was repeated and it was agreed to allow the Chinese to make the purchase. Merchant commented, "Despite the nature of the bomb and the risk it might later be used against Hong Kong or in SEA [Southeast Asia], I think we should place no obstacle in [the] path of the Chinese procurements." Assistant Secretary for Far Eastern Affairs Dean Rusk concurred.[29]

Another factor that has been noted as proof that the State Department had abandoned Taiwan by 1950 was the withdrawal of all but necessary American personnel stationed there. The State Department was preparing for a Communist invasion and did not want American personnel trapped there as prisoners of either the Communists, the Nationalists, or Taiwanese independence groups. Acting Secretary of State James E. Webb commented that the situation for American officials in Taiwan differed markedly from that which existed on the mainland, because complications that might follow a capture of American official personnel in Taiwan could be far more serious for "internal relations" than the adverse criticism that might accompany an early evacuation.[30] The State Department feared that American public reaction, particularly by the China bloc and other Nationalist supporters, to the holding of American personnel in Taiwan would preclude a negotiated policy. On May 17 orders were sent to Taibei for all staff associated with ECA construction projects and the rural development programs as well as some of consulate staff and their families to leave the island. Several days earlier the Nationalists had evacuated coastal islands, thus allowing Communist forces in the area to be concentrated solely on Taiwan; it was thus felt that the possibility of an air attack was real. Webb sent a telegram to the embassy requesting that no publicity be given to the issuance of the notices to leave because of "the bad effects which might result on Taiwan from a public statement on reduction of United States personnel."[31]

In connection with evacuation plans, the State Department asked the Philippine government on June 2 whether it would allow Chiang and top Nationalist officials to take refuge there if fighting broke out on the island. Quirino responded on June 25 that Chiang would not be welcomed: "If Chiang came to the Philippines, he would be given 24 hours to get out."[32] No further action was necessary, however, because fighting in Korea had begun and the straits of Taiwan were "neutralized" within two days.

It can be concluded that the U.S. government was preparing for a Communist invasion of Taiwan during the latter part of 1950, but it cannot be stated with certainty that the State Department was willing to relinquish the island without a fight. The outbreak of the Korean War and the stationing of the U.S. Seventh Fleet to protect the island make it impossible to determine what the American reaction would have been if an invasion of Taiwan had predated that in Korea. It is clear, however, that this action implied that the State Department perceived a connnection between Korea and Taiwan, at least concerning the goal of stopping the further spread of "Soviet-inspired" communism in Asia. Acheson has commented that the Soviet Union "had always been behind every move" in Korea and China.[33] But both South Korea and Taiwan had been excluded from Acheson's "defense perimeter" in Asia, described in his speech of January 12, 1950. There were several other similarities in the situations in Korea and Taiwan as well. The Joint Chiefs had decided by 1948 that the United States should not use its military forces to prevent either Korea or Taiwan from falling to the Communists.[34] Although South Korea was "liberated" from Japan by American troops after World War II, the American occupation came to an end on June 29, 1949, because American military strength was overextended throughout Europe and other parts of Asia. Political and economic means were to be used to create a stronger South Korea and a stronger Taiwan, independent fom the United States. But what if this were not successful? It was planned that the United States would not act unilaterally in Korea or Taiwan, but that the United Nations would take over instead. It is possible that such a plan could have been carried out in Taiwan as it was in Korea. But, as Admiral Sidney Souers, head of the National Security Council, has commented, "When [the United States] went back into Korea, the President took advantage of the situation and stuck the 7th Fleet in between Formosa and the Mainland."[35]

The outbreak of the Korean War allowed Chiang Kai-shek, who had declared himself president once again in December 1949 while Li Zongren was in the United States, to consolidate his group's position on the island. Although the State Department did not necessarily plan to save the Nationalist government, efforts were made to prevent Taiwan from falling under Communist control because of the island's strategic location.

8. Confrontation in the United Nations

The fate of the Chinese Nationalists in the United Nations and in the world community became inexorably tied to events in Korea when it became clear to the State Department that Nationalist presence on the Security Council was necessary for securing UN support for American actions in Korea. Before the outbreak of the Korean conflict, the United States was pressured to agree to seating the representatives of the People's Republic of China. The often belabored and circuitous arguments offered by the American delegation for voting against the seating of a Communist representative frustrated UN leaders and many U.S. allies. The Soviets reacted to the situation by walking out of the Security Council. These events only underscored the weakness of the fledgling international organization and the extent of American power there. They also helped to accentuate the division between the two sides in the cold war. Once it became obvious to the Truman administration that the Chinese were linked to North Korean involvement in the war, a two-China policy was solidified with the UN condemnation of the PRC. The Nationalists would represent China until 1972.

Chinese Representation Is Considered

On November 20, 1949, UN Secretary General Trygve Lie received an official communication from Foreign Minister Zhou

Enlai announcing the establishment of the People's Republic of China on October 1 and demanding that the United Nations "immediately deprive" the Nationalist delegation of its rights to represent the Chinese people there.[1] Why the PRC delayed six weeks in sending such a communication is unclear. It might have been timed to coincide with upcoming Security Council meetings, which Nationalist representative Jiang Tingfu was scheduled to chair. The Soviet Union and other Eastern European states had consistently challenged Jiang's credentials. It was also at this time that Britain and other Western states made it clear that they would recognize the PRC within a short time.

The Soviet delegate to the Security Council, Yakov A. Malik, questioned the legitimacy of the delegation headed by Jiang at a December 29 Security Council meeting to discuss the status of Kashmir. Malik and his counterpart from the Ukraine pointed out that they had informed the General Assembly of their governments' support for the communication by Zhou Enlai. Therefore, they would not regard Jiang as the representative of China or as being empowered to represent the Chinese people in the Security Council.[2]

In his reply Jiang asserted that the statement by the Soviet and Ukrainian representatives struck a blow at the legal and moral foundations of the Security Council because a minority was attempting to deny the authority of other delegations. He referred to Zhou Enlai as he "who styles himself the Foreign Minister of the so-called People's Republic of China" and said that the questions before the United Nations included "Who is Chou [Zhou] En-lai? Who made him Foreign Minister? Who created the regime called the People's Republic of China?" The Soviet representative responded that it was unnecessary "to pay attention to the irresponsible statement and slanderous inventions of a person who represents nobody in the Council."[3]

This verbal altercation prompted a State Department meeting the following day to discuss questions affecting the Nationalist government's representation on the Security Council. Ruth G. Bacon, the UN adviser to the Bureau of Far Eastern Affairs, outlined relevant American positions and summarized possible action regarding this problem. Bacon conjectured that the Soviet challenge to Jiang's credentials was intended to pave the way for questioning his right to chair the Security Council in January. If this were true, the question would arise during the first January meeting.[4]

The United States, Bacon felt, should vote against the seating of a Communist representative so long as it continued to recognize the Nationalist government. By the time the Security Council met in January, it was possible that several more states in addition to the Soviet bloc, such as Burma, India, Great Britain, Norway, Sweden, and Denmark, would have recognized the PRC. This would directly influence the Security Council voting on the seating of a Communist representative. Of the countries represented on the Council, India, Norway, the USSR, the United Kingdom, and Yugoslavia would probably vote in favor of seating the Chinese Communists, while the United States, Ecuador, Cuba, and China would vote against. It was still not clear how Egypt and France would vote, although it was thought that Egypt would probably vote against the Nationalists.[5] A lineup of six to five or six to four, with one abstention (the United Kingdom), might result unless the British persuaded the French to abstain with them.[6]

These projections were based on the assumption that the question relating to Chinese credentials would be treated as a procedural matter and therefore would not be subject to the veto. But there seemed to be good reason for arguing that approval of credentials in a case where there were two claimants to a government of a UN permanent member was more than a procedural matter. The Office of Far Eastern Affairs did not want to press for this change because the representative of the permanent member already seated could veto resolutions pertaining to his credentials indefinitely. This would only serve to create a chaotic situation. Also, it was believed that the U.S. delegation should not use its first and only veto to date on this issue.[7]

Acheson explained the administration's position on using a

veto in a message to the U.S. representative at the United Nations, Warren R. Austin, on January 5. Austin was informed that he should make it clear to Security Council members that his vote against the seating of a Communist representative did not constitute a veto "since this decision can be taken by any seven votes." If any permanent member including the Chinese representative claimed that a negative vote was a veto, the American delegate should support steps to override this action. If the United States cast the only negative vote and this was ruled a veto by the Security Council, the American delegation should request a revote and abstain.[8]

The first meeting of 1950 was scheduled for January 10. On the preceding day Jiang Tingfu met with several members of the U.S. delegation to discuss his position regarding the seating of the Chinese Communists. He said he believed that his expulsion from the Security Council or the seating of a Chinese Communist was a substantive matter that did not relate to credentials but to the question of which government was to be recognized. He planned to abstain on any motion to seat a Communist representative if seven members including the United States recognized the Communists. But if seven members not including the United States favored this action, he would use his veto to stop such a motion. Jiang added that if a motion were made challenging his right to preside over the Security Council, he planned to follow Security Council rules of procedure and turn the presidency over to the Cuban representative for the duration of the debate.[9]

The meeting on January 10 was called to discuss a resolution concerning the regulation and reduction of conventional armaments and armed forces adopted by the General Assembly in December. The first statement of the day was made by Malik of the Soviet Union. In a lengthy introduction he read the text of a letter sent by Zhou Enlai on January 8 to Secretary General Trygve Lie, President of the General Assembly Carlos Romulo, and the governments of the states represented on the Security Council informing them that the government of the People's Republic of China considered the Nationalist presence on the

Security Council illegal. Malik submitted a draft resolution that asked Security Council members to decide not to recognize the credentials of the Nationalists. Jiang, speaking as president, ruled that Malik's proposal be printed and distributed and a special meeting called for its consideration.[10]

Malik reacted to this move by stating that he objected "to any ruling given by a person who does not represent anyone here." The Soviet delegate insisted that a vote be taken on his proposal as soon as English and French texts could be prepared. Jiang called for a vote on his ruling, which was passed by a vote of eight to two with one abstention. Malik then left the Council chamber after stating that he could not participate in Security Council work until the Nationalists had been excluded from Council membership.[11]

The debate continued. The delegate from Yugoslavia, Ales Bebler, said that he agreed with the assertions made by the Soviet delegate and in his opinion the best solution was to adjourn the debate on all other issues to allow for the distribution of the Soviet resolution. He then made a formal proposal for adjournment. The American delegate, Ernest A. Gross, replied by citing Rule 17 of the provisional rules of the Security Council, which states that a Security Council representative to whose credentials objection has been made could continue to sit on the Council until the matter was decided by that body. Debate continued on the matter until the meeting was adjourned.[12]

On the following day Austin cabled Acheson asking for direction on how to deal with the situation in the Security Council. Acheson's answer reflected a desire for the State Department to maintain a relatively low profile for the time being. The American representative, for example, should not exert any influence to guide the Security Council to an early vote on the Soviet resolution even though it appeared that this tactic would probably make it impossible for the Soviets to get seven favorable votes needed for its passage. Furthermore, Austin should not express support for any alternative proposals until further study was done. He should make a statement that the Security Council must proceed with its normal business even if the Soviets refused to participate.[13]

The American assessment of the Soviet walkout, as expressed by the CIA, was that the move was designed to isolate the Chinese Communists from the Western world, particularly from the United States, and "to thwart the development of a new [American] policy in the Far East." The Soviet actions occurred at about the same time as the seizure of American consular property in Beijing. The CIA interpreted both of these tactics as having the effect of delaying American recognition.[14] "If the U.S.S.R. had intended to expedite the seating of a Chinese Communist representation in the UN," the CIA reported, "orthodox parliamentary procedures would have been more effective. Instead, the U.S.S.R. staged a dramatic boycott of UN proceedings, which actually may have delayed UN acceptance of a Peiping [Beijing] delegation. . . . The U.S.S.R. may be concerned with preventing any U.S. effort to encourage or assist the Chinese to take an independent line in dealings with Moscow. In any case, if and when the US does recognize the Chinese Communists and the Peiping representatives are seated in the UN, the U.S.S.R. will be in a position to claim that the Western Powers simply surrendered to Soviet pressure."[15]

On January 12 the Security Council reconvened. The first item on its agenda was the Soviet resolution concerning the Nationalists' credentials. Before discussion began, Jiang temporarily relinquished his role as Security Council president and asked Carlos Blanco of Cuba to preside over the discussion of the Soviet resolution.

The first to speak on the resolution was Bebler of Yugoslavia. In his statement of support, Bebler pointed out that there was no longer a majority on the Security Council that had not yet recognized the PRC—the Council was split five to five. He did not consider the sixth vote, that of Jiang, to be relevant, "for the fact that the Chiang Kai-shek government recognizes the Government of Chiang Kai-shek is of no consequence whatever."[16] The French and American delegates, Jean Chauvel and Ernest Gross, stated that they would vote against the resolution. Malik, having returned for this session, then blasted the French and the British for going along with the Americans, saying that the actions of the British in particular were hypocritical since they had already exchanged notes of recognition with the PRC. The British replied that they were waiting for a majority of UN members to recognize the Chinese Communists.[17]

The debate continued on the next day. The Ecuadorian delegate, Homero Viteri-Lafronte, and Blanco spoke against the resolution, both noting that their governments had not recognized the new Chinese government and therefore it was too early to discuss the Chinese Communists' UN representation. In the vote that followed, the resolution was rejected six votes to three, with two abstentions. Voting in favor were India, the USSR, and Yugoslavia; against were China, Cuba, Ecuador, Egypt, France, and the United States; abstaining were Norway and Great Britain. This prompted a statement by Malik in which he accused those delegates who voted against the resolution as being followers of the United States. He concluded that the Soviet Union would not recognize as legal any Security Council decision adopted with Jiang's participation; then he left the chamber.[18]

On January 18 Bebler again proposed a resolution challenging Jiang's right to preside over the Council. This proposal was rejected six to one, with two abstentions and one absent. Yugoslavia was the sole delegate in favor; India, Norway, and Britain abstained, and the USSR was absent.[19] Similar actions were taken during the following weeks in the various UN committees. Acheson sent telegrams to American delegates on those committees, informing them to vote negatively on any proposals to unseat the Nationalists or allow a Communist representative to be seated. The Soviets continued to walk out of committees that failed to approve such resolutions. On January 15, for example, the Soviet representatives refused to participate in discussions in the Economic and Social Committee, the Subcommittee on Discrimination and Minorities, and the ad hoc Committee on Statelessness.[20]

The United Nations received another communication from

Zhou Enlai on January 19. Zhou informed the secretary general and the General Assembly president that the Chinese government had appointed a UN representative. Two questions must therefore be answered: When would the "illegitimate" Nationalist delegate be expelled from the UN, and when could the "legitimate" Chinese delegate participate in the work of the UN?[21]

Lie replied to Zhou on January 20 that each body within the United Nations was competent to act on members' credentials. As a result, Chinese representation and participation would be determined by those bodies' decisions. Lie's rather weak answer did not reflect his true concern for this issue. On the same day he told reporters that "the UN's stock was at its lowest ebb as a result of the dispute over recognition of a Chinese government . . . and the work of the UN should not be made to suffer because of this 'political struggle.'"[22] The Security Council's deliberations, for example, were in jeopardy because of questions raised as a result of the Soviet refusal to participate. Several applications for membership were scheduled to be acted upon, and it was uncertain whether the Soviets would ever recognize those members who were admitted in their absence. Several delegations hesitated to act on substantive issues because of this problem.[23]

The Secretary General Criticizes the U.S. Position

On January 21 a meeting took place in Washington between Acheson, Lie, Under Secretary of State James E. Webb, Deputy Under Secretary of State Dean Rusk, Assistant Secretary of State for UN Affairs John D. Hickerson, and Assistant Secretary General of the UN Byron Price. Lie opened the meeting by stating that he was deeply concerned over the Chinese question and the Soviet walkouts from the Security Council and other UN organs. He was worried that the Soviets intended to leave the UN and pursue armed conflict. This would completely undermine his work as Secretary General. Rusk told him that he had received no information that the Soviets planned military action in the near future. Lie, apparently "visibly relieved," then added that he supposed the United Nations would go on without Soviet participation if they decided to leave the organization. Acheson commented that this was also the American view.[24]

Lie said that he understood the U.S. position in regard to the seating of the PRC's representatives and that he thought that position was "fair and reasonable." His interpretation of the American position was that "when seven members of the Security Council vote to seat a representative of the Communist regime on a procedural motion, this will be done." According to the State Department memorandum on the meeting, Lie spoke

> in high terms of the President's recent [January 5, 1950] statement on Formosa and [Acheson's] address before the Press Club [on January 12] . . . on China and Far Eastern Matters. He said that after these statements he had been encouraged and had felt that the problem was on its way to a solution. He added, however, that the seizure by the Chinese Communists of U.S., French, and Dutch official property in Peiping [Beijing], and the understandable U.S. reaction thereto, seemed to him to demonstrate that the settlement of this matter along the pattern he had previously expected will be neither easy or achieved at an early date.

Lie made it clear that he understood the American position and was not criticizing it, but he was concerned about "the whole position of the UN," especially the attitude of the Soviet Union. He noted that he had "been earnestly considering what, if anything, he could do to contribute to a solution of all these matters." He had contemplated, for example, instituting action to call a special session of the General Assembly but was not convinced that it would be helpful. Acheson agreed with Lie that such a move would not be useful. Lie then reflected whether it would be worthwhile for him to visit Moscow for discussions. He was also unsure if this would be fruitful. Acheson commented that it would not be useful.[25]

Lie met with other American representatives—UN delegates

Gross and John C. Ross in New York—a few days later to repeat his great concern over Soviet walkouts. He said that unless the problems of seating the Chinese Communist representatives were resolved within four to six weeks, he feared that the Soviets would stay out of the United Nations for good, keep the Chinese Communists out, and proceed to set up a rival organization. This would divide the world sharply in two and destroy the basic principles of unity and universality on which the UN was founded. Lie felt strongly that some method must be devised to solve these problems. Although he did not have any evidence to support his feeling that the Soviets would stay out permanently if the Chinese Communists were not seated, he had discussed the matter with Bebler to get a "Communist viewpoint," and the Yugoslav delegate had agreed. Lie also thought there might be some in Moscow who wanted the Soviet Union out of the UN, but he had found that both Soviet delegates, Malik and Konstantin Zinchenko, seemed anxious to get the question of Chinese representation settled promptly. In two conversations with Malik, Lie had wondered why the Soviets were concentrating on unseating the Nationalists rather than seating the Communists. Malik had agreed with Lie's proposal for dealing with the representation question, which involved getting the Cuban president of the Security Council or a Security Council member that had already recognized the PRC to initiate a call for a Council meeting, the purpose of which would be to hear a report from Lie guiding the Council in determining what was meant by "Republic of China" in the present circumstances. Lie would present the interpretation that the Republic of China was Communist China, and he would stress the difference between the question of recognition by individual governments and what he called recognition by the United Nations. He felt this would overcome difficulties with several Latin American countries seated on the Security Council—they could continue their policy of nonrecognition indefinitely yet still vote for Lie's interpretation. He hoped that in this way Ecuador and Cuba would be the sixth and seventh votes needed to seat the PRC.

Lie discussed this proposal with various Security Council members. Arne Sunde of Norway expressed his approval and indicated that it would not be difficult for a country that had already recognized the Communists to switch its abstention to a favorable vote on their seating. Chauvel of France thought the proposal "might not be a bad idea" but that further discussion with his government was needed. The British also needed input from home before deciding on how to vote. But Ross told Lie that this action might be premature and hasty and that it was necessary to maintain a calm attitude toward the situation.[26] Apparently nothing further beyond notifying the various Security Council delegates was done on this proposal.

The continued Soviet refusal to participate in UN activities paralyzed the organization during the early months of 1950. Most delegates hesitated to work on substantive issues without the Soviet presence, and time was spent in arguments between the Yugoslav and Chinese delegates over Security Council action on Zhou Enlai's communications. This was clearly contrary to the State Department's goals, and Acheson instructed the American UN mission to stress the U.S. view that "no UN member can by its willful absence impair normal functioning of any UN organ or validity of decisions which it may take." The United States supported the maintenance of the normal range and tempo of activities in all UN bodies.[27]

Lie took exception to the U.S. position. In several meetings with American representatives at the UN, he criticized the American role in attempting to prevent the seating of the Communists. He had learned, for example, that the American ambassador in Ecuador had been instructed to urge that government to hold off on the representation question. Lie said that he did not view such action as essential to American policy but that it hindered his own work.[28] He indicated that the Americans had made a fundamental mistake in their China policy five years earlier and had perpetuated that mistake ever since. Furthermore, he did not feel that the American policy toward Chinese recognition necessitated using pressure on other governments to prevent the seating of the Communists in the United Nations. He hoped that the United States "would let nature take its course, nature," according to Ross, "being his own efforts to get enough votes in the Security Council to seat the Chinese Communists."[29]

The United States Adopts a Position of "Neutrality"

Lie's campaign to secure the PRC's position in the United Nations and his criticism of U.S. actions forced the State Department to retreat, at least temporarily. Gross informed Acheson that Lie's persistent efforts threatened to cause confusion within the international organization and gave a distorted view of the American position. It was undesirable to create the impression that the American delegation and the secretary general were consistently in conflict. Gross suggested that although the United States should not change its voting pattern on this question, it should make it clear to other delegations that it was not pressuring them to vote in a certain way. Gross acknowledged that Lie's discussion of the Ecuadorian case was particularly sensitive because that embassy had informed the State Department on January 12, 1950, that it intended to break diplomatic relations with the Nationalists but delay recognition of the PRC. The State Department had replied on January 17 that this would have an "important effect upon [the] voting situation, already delicate, in [the] S[ecurity] C[ouncil]" and the "Ecuadorian Government might wish to consider deferring, at least for [the] present, breaking relations with [the] Nationalist government." After lengthy conversations with Ecuadorian ambassador Viteri-Lafronte, Gross concluded that these statements had resulted in a dilemma for the State Department. The Ecuadorian government's position was that the American statements amounted to a request that it must continue to honor if it did not want to "take the risk of causing [the] U.S. harmful consequences which it would not under any circumstances want to do." It was concluded to be in Ecuador's best interests to take action that was of the greatest benefit to the United States. The State Department was being criticized for having made such a "request," yet it did not want to be put in a position where another communication could be interpreted as a release from this request, which could then be construed as an American preference for an Ecuadorian vote in favor of seating the Communist representatives. It appeared to Gross that Lie was "busy creating his impression."[30]

To remedy this situation, Gross suggested that a statement clarifying the American position be sent to American embassies in UN member states. He wanted the State Department to emphasize that the U.S. government did not seek "to bring pressure" or otherwise encourage, discourage, or influence other delegations in their vote on the Chinese representation question.[31] On March 23 Acheson sent such a communication to fifty-six embassies, summarizing American policy and noting that it was important for consular officers to emphasize that UN problems arose not from the American attitude but from Soviet unwillingness to comply with a majority decision.[32] Thus, the U.S. government undertook what the American delegation to the UN considered to be a position of "neutrality" on the question of Chinese representation.[33] But this policy changed after hostilities broke out in Korea on June 25. The United States needed support for actions taken in Korea under UN jurisdiction, and it became important to keep the Chinese Nationalists seated on the Security Council.

On the day the fighting commenced, Ambassador Gross outlined the developments in Korea for the Security Council and proposed that the United Nations consider "this wholly illegal and unprovoked attack by North Korean forces" as a "breach of the peace and an act of aggression."[34] On a proposal by the American delegate, a representative from South Korea, not yet a UN member,[35] was invited to sit on the Security Council during consideration of the question. Gross then presented a resolution calling on North Korean authorities to cease hostilities and withdraw their forces to the border and asking all UN members to render assistance. The Cuban, Chinese, French, British, and Ecuadorian delegations voiced support for this motion. The resolution was approved by a vote of nine to one, with Yugoslavia casting the negative vote and the USSR absent. The Yugoslav representative then submitted another resolution which called for the immediate cessation of hostilities and invited the North Korean government to state its case before the Security Council. This was rejected by a vote of six to one, with three abstentions.[36]

Two days later the Council heard from the UN Commission on Korea[37] that North Korean authorities were carrying out a "well-planned, concerted, and full-scale invasion of South Korea." Gross submitted another resolution asking the Security Council to note that North Korean authorities had not complied with the June 25 resolution and that urgent military measures were required to restore peace. It recommended that UN members furnish such assistance to South Korea to repel the North Korean attack. In support of this resolution, Gross read a statement by President Truman made earlier that day announcing that he had ordered United States air and sea forces to give cover and support to South Korean troops and the Seventh Fleet to prevent an attack on Taiwan by the Chinese Communists, considered by the Americans to be backers of the Communist expansion in Korea. Truman had also called upon the Chinese Nationalists to cease air and sea operations against the mainland and had accelerated military assistance to the newly independent Philippines and French Indochina.[38]

The Yugoslav delegate, Bebler, proposed an alternative resolution asking the Security Council to renew its call for the cessation of hostilities and invite a representative from North Korea to participate in UN mediation. He noted that Korea and the Korean people were "victims of 'spheres of influence'" and that the Security Council, after only two days of fighting, should not abandon hope that Koreans would negotiate in their own interests. The Council rejected this resolution by seven votes to one (Yugoslavia), with one member absent (USSR) and two not participating in the vote (Egypt and India). The U.S. resolution was adopted by an identical voting pattern.[39]

Policies Toward Korea and Taiwan Become Entangled

Although support for the American resolution in the Security Council was sufficient for the State Department to receive UN backing for its Korean policy, the close vote, whereby only the minimum of seven votes needed to pass the resolution was achieved, reaffirmed the importance of the presence of the Nationalists on the Security Council. This view was specifically expressed within the State Department in two memoranda written on June 29 by UN advisers G. Hayden Raynor and Ruth Bacon. Raynor explained that because of the need for seven votes for "quick emergency action" in Korea, the U.S. position of "neutrality" concerning the Chinese representation question should be changed. The situation, moreover, had "an urgent aspect" because the American delegation had received word before the outbreak of fighting in Korea that the British might change their vote from one of abstention to favoring the unseating of the Nationalists, and their position was still uncertain. Raynor concluded that "the situation calls for us to urge the British not to make this change . . . and I would also think serious consideration should be given to our letting other delegations generally know that because of recent developments we think, at least at this time, there should be no change in the status quo on the representation question."[40] Bacon added that "under existing circumstances [the] need for a dependable majority in the S[ecurity] C[ouncil] would appear to be the overriding consideration. Accordingly, it is suggested that we should inform other friendly powers that for the present we believe that any change in Chinese representation would be undesirable."[41] On July 3 Acheson repeated these views to Austin at the UN and said that if Austin received evidence that the Chinese representation question might be raised, he should express the view to other delegations that the United States did not want the issue considered.[42]

On the same day, the secretary general of the Indian Ministry for External Affairs and UN representative, Girja Bajpai, informed the State Department that the Indian government was of the opinion that it was important for world peace that the PRC and the Soviet Union sit on the Security Council and was therefore trying to persuade other members to vote for the immediate admission of the Chinese Communists. Acheson cabled the American embassy in India and the UN delegation, telling them to emphasize to the Indians that the United States considered it undesirable for the Chinese question to be brought before any UN organ while the Korean situation had not been resolved. A Chinese Communist representative on the Security Council would bring "serious and effective obstruction" in this case, according to Acheson.[43] Two personal messages to Acheson from Prime Minister Jawaharlal Nehru expressing the Indian position failed to convince the State Department to change its views.

On July 10 the Indian ambassador at Moscow spelled out an Indian plan in a "secret personal" letter delivered to American Ambassador Alan G. Kirk for Acheson. The communication read:

> I [Indian Ambassador Sarvepalli Radhakrishnan] have been thinking a great deal about [the] Korean situation. I know that anything we do should not appear as [a] matter of appeasement. What do you think of [a] settlement on these lines: (1) that America support the admission of the People's Repulic of China into S[ecurity] C[ouncil] and UN; (2) and that [the] SC with China and U.S.S.R. on it support immediate cease fire in Korea and withdrawal of North Korean troops to 38th parallel and mediation by UN for creation of united, independent Korea?
>
> Postscript: If we are able [to] bring main disputants into SC to consider outstanding questions it may well be beginning of new chapter.[44]

The Indian government also sent similar messages to authorities in Moscow and Beijing, thus beginning several days of "informal mediation" to resolve the impasse in the Security Council. Although the Soviets made it clear that they would not accept the second point in the formula, the Chinese said that they agreed to both. The Indian government viewed this divergence between the Soviet and Chinese Communists as "most significant" and emphasized that American acceptance of the formula would bring about a split between the Kremlin and Beijing. This would spotlight Soviet unwillingness to cooperate with the United States in a peaceful settlement to resolve the situation. Furthermore, positive support by the United States for Chinese Communist representation was not essential: "Abstention, coupled with [a] friendly word to Ecuador and Cuba, would probably do just as well."[45]

According to Acheson, when Truman read the Indian letter at a meeting on July 10, its contents had disturbed him. The president was then assured that work was under way in the State Department to deal with the matter.[46] The Department handled the issue in two ways. During the following week a letter, written to be released to the press, was sent to Nehru stating that the U.S. government would not agree to either point in the formula.[47] In the meantime, Ambassador Kirk in Moscow attempted to persuade Radhakrishnan to cease his mediation efforts. Kirk commented that the Indian ambassador accepted the view that the United States government would not back down on the Chinese representation question but hoped that the Americans would eventually recognize the PRC.[48] At the United Nations the Indian delegation continued to support the Chinese Communists in their bid for representation.

The matter came to the forefront again when the representative of the Soviet Union assumed the presidency of the Security Council on August 1. Malik notified the secretary general on July 27 of his intention to return to the Council for his turn as presiding officer. This stimulated activity among the American, British, French, Norwegian, and Egyptian delegations to formulate a plan of action if the Soviets raised the question of Chinese representation. Acheson notified the UN delegation as well as the embassies in these four countries that any Soviet attempt to rule that China was improperly represented should be opposed on the basis that the Security Council itself was to decide on that question and the president had no authority to make such a ruling. Acheson also desired that a representative from a state already recognizing the PRC should challenge Malik's authority. Truman approved this plan on July 31 and agreement was reached with the French, Norwegian, and Egyptian delegations. The British refused to cooperate.[49]

Malik opened the August Security Council meetings with a ruling that the Chinese delegate did not represent China and therefore could not participate in Council meetings. Gross then challenged the ruling, stating that "no President had the authority to rule, by arbitrary fiat, upon the status of the representation of a [member] country." A vote was taken and Malik was overruled by eight votes to three (India, USSR, Yugoslavia). Malik countered that this decision was illegal because the person concerned was the spokesperson of a group that represented no one.[50] On August 3 Malik proposed to include in the agenda an item entitled "Recognition of the Representative of the People's Republic of China as the Representative of China." This resolution was defeated by a vote of five to five, with one abstention.[51]

The American determination to keep the Chinese representation question off of the Security Council agenda subsequently led to problems. Acheson met with Truman on August 3 to review "the difficult situation" in which the U.S. government found itself with its allies in the Security Council "by reason of the Russian ability to play on the Korean situation, Formosa, and the Chinese Communists." Acheson acknowledged that "by keeping these matters connected they [the Soviets] could mobilize certain nations in opposition to [the United States], some on each issue."[52] The British, for example, continued to insist that the Chinese representation question was unrelated to the Korean issue, and that they desired to seat the Communists. The Indians felt that Chinese Communist representation would be beneficial to East-West cooperation and to a resolution of the Korean War. They would persist in efforts to support the Chinese Communists.[53]

Acheson also pointed out to the president "the great need for circumspection in regard to Formosa[54] and the importance of not having the Communists seating issue arise for a vote. . . . To seat the Chinese over our objections," he continued, "would whip up opinion here against our Allies. We could not meet the views of our Allies [who favor Chinese Communist representation] as long as the fighting in Korea continued." Acheson also noted that talks with the British would begin in order to convince them that the preservation of unity among the allies was of the greatest importance.[55] This situation also caused the State Department to withhold information from the secretary general. For example, on January 29 Ambassador Koo told Livingston Merchant, then deputy assistant secretary of state, that the Nationalist government was willing to send 33,000 troops to South Korea to be commanded by General Douglas MacArthur. Merchant expressed his concern that Koo not relate these intentions to Lie.[56]

The General Assembly opened on September 19 with Benegal N. Rau, delegate from India, circulating a draft resolution calling for the seating of the Chinese Communists. This was an unusual move in the organization's short history since it was not customary to deal with a resolution at so early a stage, before the delegates' opening speeches. An acriminious debate followed. The Soviet delegate, Andrey Y. Vyshinsky, proposed two other resolutions, calling on the General Assembly to decide that the Nationalists could not take part in the organization because they did not represent China and to invite the representatives of the PRC to take part in UN bodies. In Acheson's address to the delegates he asked why "the Assembly should eject from representation here representatives of the Government of China which has participated in the founding of the UN and has represented China ever since and that the Assembly should substitute for that representation the representatives of another regime in China."[57]

Vyshinsky began his reply by citing evidence against the Nationalists published in the White Paper. He also referred to Acheson's January Press Club speech: "I should like to remind Mr. Acheson of a statement he made not so long ago, on 12 January 1950 to be exact, in which he recognized that the Chinese people had completely withdrawn its support from the Government. I should like to remind him that earlier still—about a year ago, in August 1949—in his letter of transmittal of the famous White Paper which came out at that time, he said about Kuomintang [Nationalist] China: 'In the opinion of many observers they had sunk into corruption, into a scramble for power.' "[58]

On the following day Acheson replied to these statements by declaring that "in the opinion of the United States the international community has a legitimate interest in having the Formosan question settled by peaceful means." He then proposed that the General Assembly "direct its attention to the solution of this problem under circumstances in which all interested parties shall have [an] opportunity to express their views and under which all concerned parties will agree to refrain from the use of force while a peaceful and equitable solution is sought."[59]

Meanwhile, the American and British delegations had been working on a resolution that would establish a Special Committee of the General Assembly to study the question of Chinese representation. The Americans made it clear, however, that they did not want such a committee to begin consideration of this question until late in the 1950–51 General Assembly session.[60]

When the votes on the motions to seat the Chinese Communists were carried out, the Indian resolution was rejected by a vote of thirty-three to sixteen, and the Soviet resolutions were also rejected, by a vote thirty-eight to ten, with eight abstentions.[61] The Canadian delegate then proposed a resolution asking the Assembly to establish the special committee formulated by the American and British delegations, to consist of the Assembly president and six other representatives selected by the president to consider the question of Chinese representation. This proposal was passed by thirty-eight votes to six, with eleven abstentions. A second part of this resolution, which allowed the Nationalists to retain their seat in the UN pending the committee's decision, was also passed. The committee was established on December 12, 1950. Representatives from Canada, Ecuador, India, Iraq, Mexico, the Philippines, and Poland were selected to serve, but work did not commence until the fall of 1951 because the American delegation pushed for further postponement of consideration of the Chinese representation question after the Chinese Communist intervention in the Korean War in November 1950.[62]

Conclusion

The outbreak of hostilities in Korea caused the United States government to act to prevent the seating of PRC representatives. The ability of the United States to control the voting in UN bodies, particularly the Security Council, through "discussions" with allies made it impossible for the Soviet-bloc states and other Chinese Communist supporters, such as India and Great Britain, to achieve their goals of replacing the Nationalist representatives. Moreover, after the entrance of Chinese Communist troops into the Korean War in November 1950 and the subsequent UN vote to condemn the PRC as an aggressor on February 1, 1951, the stage was set to keep the Chinese Communists out of the organization for over twenty years.

Although to some it may appear that this was merely a culmination of an American plan to prevent the PRC from participating in the United Nations, this was not the case. After January 1950 pressure on the United States from its allies as well as from UN Secretary General Trygve Lie forced the State Department to acquiesce to the apparent inevitability of Chinese Communist participation in the UN. The Soviet boycott of UN agencies coupled with refusal by Soviet allies to discuss other issues without first resolving the Chinese representation question led Secretary General Lie and delegations friendly to the United States to try to convince the Americans that their position was wrong. By March 1950 the United States government had adopted a position of "neutrality" on this issue, and telegrams were sent to embassies throughout the world stating that the Americans would no longer interfere with the decisions of other delegations on the representation question. One can only conjecture whether the State Department felt that its allies would not change their votes or whether it was ready to be outvoted in the Security Council.

This attitude changed with the outbreak of the Korean War when the presence of the Nationalists on the Security Council became crucial for American action in Korea. The State Department then renewed its efforts at influencing other delegates to comply with its views. This policy proved successful. Some American allies, particularly those in Latin America, voted with the Americans against the seating of the Communists while others abstained.

The debates within UN bodies on the Chinese representation question revealed the underlying weakness of the organization. With each nation pursuing its national interests and attempting to garner support from its allies, this problem prevented effective work on other matters from taking place and further contributed to cold war tensions. Furthermore, it demonstrated the amount of control the United States enjoyed over the organization at the time, as American policy toward the PRC held sway in the United Nations.

9. Conclusion

As a Communist victory in China became more certain by 1948, the Truman administration was beset with questions concerning the development of a new American China policy. The decision to grant additional aid to the failing Nationalist government with the China Aid Act of April 1948 came after a series of debates within the State Department and Congress. The form in which the initial aid proposal was offered to Congress was put forth by Secretary of State George Marshall, who refused to heed the advice of his advisers in the State Department's Office of Far Eastern Affairs, the National Security Council, or the embassy in China for a more comprehensive assistance package. Marshall had become discouraged with the corruption and lack of leadership associated with Nationalist President Chiang Kai-shek, and he did not want the United States to "go down to defeat" with the Nationalists.[1] The assistance program he formulated was one of limited economic aid designed to retard the inevitable deterioration of the Chinese economy.

When the China aid proposal went before Congress, it received a military aid proviso and was linked with similar programs directed toward Europe in the House Committee on Foreign Affairs. Nationalist supporters in Congress, known as the China bloc, pushed for even more money and an increased American commitment to the Chinese government, but their efforts were unsuccessful. The Senate Foreign Relations Committee agreed with Marshall, and the proposal was further revised to exclude assistance earmarked specifically for military purchases. A grant of $125 million allocated for whatever purposes the Nationalist government deemed necessary replaced the military aid. On the one hand, the Truman administration would not support a program that implied a commitment to keep China from becoming Communist as was done for Greece and Turkey in the postwar European Recovery Program. On the other, Nationalist supporters in the United States were appeased by this type of aid. The resulting package was a bill for limited assistance that could possibly postpone what was considered to be an inevitable Communist victory.

The passage of the China Aid Act by Congress on April 3, 1948, did not put an end to the problems concerning these funds. Implementation of the program was difficult because of disagreements between American and Chinese administrators over who should control the disbursement of funds. The newly formed Economic Cooperation Administration in Washington was put in charge of the program, but there were continuous conflicts with the Nationalists over the methods of implementation. The Chinese would accept no conditions that impaired their sovereignty, and the Americans wanted assurances that the funds were not hoarded or squandered by corrupt Nationalist officials.

Meanwhile, Communist armies advanced southward from positions captured in North China, making the Nationalists' situation look bleak. In November 1948 American Ambassador John L. Stuart, who was at the Nationalist capital in Nanjing, predicted that the Communists could take that city within two weeks.[2] A reappraisal of the American aid program for China was necessary. In the same month Acting Secretary of State Robert A. Lovett requested the Joint Chiefs of Staff to appraise the strategic implications for U.S. security if a Communist government were to take over the island of Taiwan. The Joint Chiefs determined that the results would be "seriously unfavorable" because an unfriendly administration on Taiwan would have the potential of dominating adjacent sea routes, thereby threatening American interests in Japan, the Philippines, the Ryukyus, and the Malay peninsula.[3] The U.S. government thus planned to utilize "all possible political and economic means" to keep Taiwan "out of the hands of the Chinese Communists."[4] This policy evolved and was strengthened throughout 1949 and early 1950 and reached its conclusion on June 27, 1950, when military means were used to keep the Chinese Communists from invading the island.

Throughout 1949 the Truman administration dealt with the uncertain situation in China in an attempt to protect American interests in East Asia. During the early part of that year, it was still unclear which government the Communists would eventually have to battle in order to take control of Taiwan. American personnel on the island were to investigate the viability of Taiwanese independence groups as well as determine potential Nationalist strength. The possibility of using the United Nations as a forum to push for independence from China was also considered. Meanwhile, aid from the 1948 China Aid Act and its 1949 amendments continued to pour onto Taiwan.

By June 1949 it was clear that the Nationalist government intended to make the island its last stronghold. Former President Chiang Kai-shek, having resigned his post in January in favor of Li Zongren, who remained on the mainland, moved to the island, taking with him most of the Nationalist army and much of the revenue from the government's treasury. Moreover, the Nationalists realized by this time that the U.S. government was interested in keeping Taiwan from the Communists.[5] Although the State Department initially did not intend to save the Nationalist government, the strategic importance of Taiwan made it impossible to separate assistance for the island from aid to Chiang Kai-shek after mid-1949.

Once the State Department was resigned to the reality of the situation on Taiwan, a policy evolved whereby the Americans tried to maneuver to keep Taiwan free from communism while seeking relations with the Communist authorities, who were rapidly consolidating their position on the mainland. This was done in several ways. On August 5, 1949, the State Department released the China White Paper, which gave detailed documentation of American aid to the Nationalists since 1937, outlined the corruption, incompetence, and other problems that plagued that government, and announced that the Truman administration would not give the Nationalists any additional aid to stop the Communists from capturing the remainder of the Chinese mainland.[6] Truman's insistence that the volume be released before the total collapse of the Nationalist government on the mainland, despite arguments against such a move by his advisers, was useful for American policy toward both Taiwan and the Chinese mainland. An American goal was to push the Nationalists to strengthen themselves and become more independent on the island with assistance already allocated so that they might ward off a Communist threat without American military aid. On the other hand, the denial of an association with Taiwan would show the Communists that the United States had no designs on the island. This move might have opened the way for discussions between the Americans and the Chinese Communists if Communist intelligence had not correctly realized that American aid to Chiang on Taiwan continued.

In January 1950 Truman and Secretary of State Acheson made statements that supported the dual goals for American policy toward Taiwan and the mainland. On January 5, in a speech to the American people, the president announced that the United States had "no predatory designs" on Taiwan and would not provide military aid or advice to Chinese forces on the island because Nationalist resources were adequate for its defense. One week later Acheson, in an address before the National Press Club, summarized U.S. policy toward East Asia and outlined a "sphere of strategic concern" that excluded both Taiwan and South Korea. The American public, particularly the China bloc, assessed these announcements as a plan to "abandon" the Nationalists on Taiwan. But American Ambassador to Korea John J. Muccio has claimed that this was not the correct interpretation. According to Muccio, Acheson implied that the U.S. government would aid those areas excluded from his "defensive perimeter" through the

United Nations and not with unilateral action. Therefore, this did not necessarily signal the future neglect of Taiwan or South Korea. Moreover, military aid to Taiwan, including napalm bombs and other incendiaries, continued despite the assurances in Truman's speech. Since the Chinese Communists were well aware of this situation, Truman's and Acheson's statements had little impact on the improvement of Chinese-American relations.

One way in which the United States had tried to keep a toe-hold in China was through economic relations. Trade between American oil companies and the Communists continued until after the outbreak of the Korean War, and American personnel at the American-owned Shanghai Power Company stayed with that firm until December 1950 when the Chinese Communists took over the plant. The objective of encouraging economic ties with the Chinese Communists was spelled out as early as July 1948. It was hoped that this would help to separate the Chinese from Soviet influence.[9]

The question of recognition of the PRC was a problem for the Truman administration. American economic ties with the Communists had been kept secret from Congress and the public because some in Congress and elsewhere were pressuring the administration to continue recognizing the Nationalists as the de jure government of China. Republicans in particular stepped up their campaign in time for the 1950 congressional elections; the China issue became a focal point for China-bloc Republicans after Truman's surprise electoral victory in the 1948 election.[10]

Although this was perhaps one factor that governed Truman's failure to accord recognition to the Communists, the reason given by Secretary of State Acheson for the delay was the situation in Mukden, where the American consulate staff had been imprisoned by the Communists from November 1948 to December 1949. Acheson has commented that "nobody could think about" recognition because of this matter.[11] Efforts to resolve this problem throughout 1949 only led to greater hostility and deeper tensions between the Americans and the Chinese Communists. Both sides resisted approaches by the other to work out an accommodation. Several other incidents against Americans in China prompted Truman to assert in December 1949 that if Americans were not "treated respectfully," there was "no point in recognizing" the PRC.[12] As a result, when the Americans from Mukden finally arrived home later that month, Chinese-American relations had deteriorated to the point where agreement seemed unlikely.

During the early months of 1950, the State Department continued efforts to hold discussions with Communist authorities while the Chinese tried to pressure the Americans into hastening recognition. One such move by the Chinese was a threat to requisition parts of foreign consular property in Beijing in January. In reply, the U.S. government announced that it was prepared to close all its establishments and withdraw all personnel from China. By contrast, the British government granted recognition to the PRC within a few days. The threat was carried out and one building of the American consular compound was taken over by Chinese authorities, while the British were left alone. On January 14 the State Department officially announced its intention to withdraw its embassy from China.[13]

The basic issue for the Truman administration was the failure of the Communist authorities to accord American personnel diplomatic immunity in China. On the other hand, the underlying problem for the Chinese Communists was the de jure recognition of the Nationalists on Taiwan by the United States coupled with continued economic and military aid to that group after the founding of the PRC on October 1, 1949. Moreover, after the Nationalists began bombing congested mainland cities with American equipment in February 1950, the Communists blamed the United States for its collaboration in the killing of innocent people.[14] Trade with a few American companies was not enough for the new Chinese authorities; the Communists wanted recognition as well as assurances that the United States would not interfere with a takeover of Taiwan. Neither side would give on these issues.

The British also were disturbed at American support for the Nationalists because attacks against mainland cities and Communist trade inevitably were focused on British-owned ships, factories, and residences. Anglo-American relations were strained over the differences in the two allies' China policies. Despite attempts by Acheson to win British support for a common policy toward the Chinese Communists, the British began to deviate as early as March 1949. They had succumbed to Communist pressure, which threatened their economic interests, and therefore accorded the new authorities in North China de facto recognition even before the PRC was formally established.[15] From this point, the evolution of British and American policies toward China and Taiwan has little similarity. The tension reached a high point at a March 1950 meeting between Acheson and British Ambassador Oliver Franks in which Franks accused Americans of aiding Nationalist bombing of British property while Acheson claimed the British were supporting the Communists against the Nationalists and Americans on Taiwan.[16] Acheson later commented that one reason he backed American companies in their trade relations with the Communists, in spite of the souring of diplomatic relations, was that he did not want to see the British gaining more trade from American losses.[17]

These problems—concerning the recognition of the PRC and British policy toward China—were also carried into the United Nations. In October 1949 the U.S. government made it clear that it did not want the Nationalist representatives replaced by those of the PRC until the United States granted the latter recognition. American influence within the international organization allowed for the postponing of the seating of the Communists during the General Assembly session that met from September to December 1949, but by early 1950, after the PRC had been recognized by twenty-four nations, including Great Britain, this position became increasingly unpopular. Representatives from the Soviet Union began a boycott of UN agencies in January to protest the presence of the Nationalist delegation, and Soviet allies refused to discuss other issues unless the Chinese representation question was resolved. As a result, the efficacy of the United Nations was seriously challenged. Secretary General Trygve Lie feared that the Communist bloc states would leave and set up a rival organization. In several discussions with Acheson and members of the American delegation, Lie criticized American pressure on its allies to vote against the seating of the Communists. He felt that Chinese Communist participation was necessary to promote world peace. Because of these disagreements, by March 1950 the State Department's attitude appeared to soften, and Acheson cabled American embassies to notify diplomatic personnel that they should no longer attempt to influence other delegations' decisions on the representation question. They were to make it clear, however, that the United States still recognized the Nationalists.[18] With the outbreak of the Korean War on June 25, 1950, however, the Americans again urged other delegations to join them against seating the Communists.

The importance of the Nationalist delegate's presence on the Security Council was demonstrated to the State Department during the first few votes taken on the Korean situation. Seven votes were needed for the United Nations to take action, and that minimum was the total number of positive votes the United States could garner. The Nationalists consistently voted with the United States, so they had become indispensable allies. Therefore, American advisers suggested a change in the policy of "neutrality" on the question of representation, as spelled out in the March telegram, and Acheson agreed. Accordingly, on July 3 the U.S. delegation was informed that it should discourage other delegates from raising the representation question.[19]

Some U.S. allies, particularly Britain and India, disagreed with this stand. Neither perceived the connection between the Korean War and Chinese representation question in the same way as did the Americans. The Indian delegation, furthermore, worked to seat the PRC representatives despite the American attitude. But efforts to avoid a vote on this question continued to be successful. After the intervention of the Chinese in the Korean War in November 1950 and the UN vote to condemn the PRC as an aggressor in February 1951, the Nationalists on Taiwan remained as the representatives of China in the UN until 1971.

The outbreak of the Korean War was indeed a turning point for United States-China relations, but not for all aspects of American policy toward East Asia. Plans for securing Taiwan from Communist takeover had been worked out well before the Seventh Fleet was placed in the Taiwan Straits, but unilateral military action was not considered a wise move by the Truman administration until June 25, 1950. American military strength was already overextended in Europe and other parts of Asia, and there was no desire further to stimulate Chinese irredentist feelings over American actions in Taiwan. Nevertheless, the administration found it necessary for American security in East Asia to take military action toward China after the North Koreans moved south. The perception of a connection among the North Korean, Chinese, and Soviet Communists was behind such a move.

This link between Asian and Soviet Communists was noted long before the Korean War. The Soviet Union was considered to be the force behind all Communist movements in Asia as well as Europe. The prompt Soviet recognition of the PRC on the day following its founding "left no doubt that the Chinese Communists [were] solidly aligned with the Kremlin."[20] On June 7, 1950, Assistant Chief of Staff Major General C. A. Willoughby outlined American security concerns in East Asia for the State Department. It was concluded that the "sharp orientation" of the PRC toward the Soviet Union made events in China "internationally significant." From the viewpoint of the General Staff, "the immediate military factor [was] the expansion of Soviet capabilities from Siberia to south China and the resulting use of Chinese bases, including Formosa, as they affect the American defense line: Japan-Okinawa-Philippines. . . . As the war in China swe[pt] south toward the borders of French Indo-China and Burma, there [was] an increasing threat to all of South East Asia." Military penetration of Southeast Asia by Communist forces was considered "a real possibility."[21]

Once the fighting on the Korean peninsula began, American policy toward East Asia was reviewed. In discussing this connection between the events in Korea and other areas, the president and his advisers determined that "the invasion of Southern Korea cannot be regarded as any isolated incident. It alters strategic realities of the area and is a clear indication of the pattern of aggression under a general international Communist plan." It "seem[ed] clear" to the administration "that Formosa now offer[ed] an opportunity for United States action of general utility in maintaining the peace of the Pacific area." The recommendation to the president was "that as soon as North Korean noncompliance with the Security Council resolution [ordering a cease-fire was] known, the 7th Fleet be ordered to proceed to Formosan waters."[22]

The action on the Korean peninsula clearly was perceived as the beginning of further Communist expansion not only in Asia but throughout the world. On the day following the commencement of fighting, Truman discussed this situation with his administrative assistant, George M. Elsey. The president expressed concern not just for Taiwan but for other parts of the world. "Korea," he said, "is the Greece of the Far East. If we are tough enough now, if we stand up to them [the Soviets] like we did in Greece three years ago, they won't take any next steps. But if we just stand by, they'll move into Iran and they'll take over the whole Middle East. There's no telling what they'll do, if we don't put up a fight now."[23] Chinese-American relations thus became associated with fears of Soviet expansion and with a hard-line policy toward the Soviets that was solidified in 1950. American protection of the Nationalists on Taiwan made the normalization of relations between the PRC and the United States impossible, a situation that was not resolved for nearly thirty years.

Notes

Chapter 1

1. Ernest R. May, *The Truman Administration and China, 1945-1949* (Philadelphia: Lippincott, 1975), p. 3.
2. "Meeting with the President," 17 October 1949, 893.01/10-1949, National Archives, Washington, D.C.
3. Dean Acheson, "United States Interests in China," undated (late 1949), Papers of Harry S. Truman, President's Secretary's Files, Harry S. Truman Library, Independence, Missouri.
4. Memorandum of Conversation by Marshall, 11 June 1948, in U.S. Department of State, *Foreign Relations of the United States* (Washington, D.C.: GPO, 1973), 1948, 8: 91-99; hereafter cited as *FRUS*.
5. See, for example, Ralph N. Clough, *Island China* (Cambridge: Harvard University Press, 1978), pp. 6-7; "Editorial Note," *FRUS*, 1950, 6:367.
6. National Security Council, "Current Position of the United States with Respect to Formosa," 4 August 1949, *FRUS*, 1949, 9:369-71.
7. Dean Acheson, "The Responsibility for Decision in Foreign Policy," in Acheson, *This Vast External Realm* (New York: Norton, 1973), p. 201. This quotation specifically referred to decisions regarding Korea in June 1950, but it also seems appropriate to events concerning the recognition question.
8. See Dean Acheson, Princeton Seminars, 22-23 July 1953, Papers of Dean Acheson, Truman Library.
9. Cabinet Meeting, 22 December 1949, Papers of Matthew J. Connelly, Truman Library.
10. Clubb to Acheson, 10 January 1950, *FRUS*, 1950, 6:273-75.
11. Memorandum by Rusk to Acheson, 14 April 1950, *FRUS*, 1950, 6:327-28; Nancy Bernkopf Tucker, "Nationalist China's Decline," in *Uncertain Years: Chinese-American Relations, 1947-1950*, ed. Dorothy Borg and Waldo Heinrichs (New York: Columbia University Press, 1980), p. 162.

12. "Blockade of China and the Hong Kong Aircraft Problem," 27 March 1950, Acheson Papers, Truman Library.

13. See Warren W. Tozer, "Last Bridge to China: The Shanghai Power Company, the Truman Administration and the Chinese Communists," *Diplomatic History* 1, 1 (Winter 1977):66.

14. National Security Council, NSC 41, 28 February 1949, *FRUS*, 1949, 9:826-34.

15. "Points Requiring Presidential Decision," undated, Papers of George M. Elsey, Truman Library; U.S. Department of State, *Bulletin*, 3 July 1950, p. 5.

16. U.S. Department of State, *Weekly Review*, 23 November 1949, Elsey Papers, Truman Library.

17. Kennan to Acheson, 6 January 1950, *FRUS*, 1950, 1:127-38.

18. 611.93/3-150, 1 March 1950, State Department, Washington, D.C.

19. Record of an Interdepartmental Meeting on the Far East at the Department of State, 11 May 1950, *FRUS*, 1950, 6:90-92.

Chapter 2

1. For a detailed account of aid to China from 1937 to 1950, see "Summary of U.S. Government Economic and Military Aid Authorized for China Since 1937," undated, Truman Papers, Official File, Truman Library.

2. Representative Walter H. Judd, *Congressional Record* 93 (Part 9), 80th Congress, 1st Session, 1947, p. 11037.

3. Ambassador Philip D. Sprouse, Chief, Division of Chinese Affairs, 1948, Oral History Interview, Truman Library, p. 35.

4. See, for example, Ross Y. Koen, *The China Lobby in American Politics* (New York: Macmillan, 1976), pp. 198-200; Lewis M. Purifoy, *Harry S. Truman's China Policy: McCarthyism and the Diplomacy of Hysteria* (New York: New Viewpoints, 1976), pp. 66-69; Tang Tsou, *America's Failure in China, 1941-1950* (Chicago: University of Chicago Press, 1963), pp. 441-93.

5. Memorandum of Conversation by Marshall, 11 June 1948, *FRUS*, 1948, 8:91-99, also cited in May, *The Truman Administration and China*, p. 93.

6. "Legislative History of the China Aid Act," 19 April 1948, Papers of John D. Sumner, Truman Library, p. 1.

7. Tsou, *America's Failure in China*, p. 465.

8. "Legislative History of the China Act," p. 1.

9. "Editorial Note," *FRUS*, 1948, 8:442.

10. Ibid.

11. Cabot to Butterworth, 6 February, 1948, *FRUS*, 1948, 8:467-71.

12. Stuart to Marshall, 29 January 1948, *FRUS*, 1948, 8:464.

13. Stuart to Marshall, 1 March 1948, *FRUS*, 1948, 8:364.
14. Butterworth to Thorp, 30 December 1947, *FRUS*, 1948, 8:443.
15. See Tsou, *America's Failure in China*, pp. 462-65.
16. Butterworth to Lovett, 21 January 1948, *FRUS*, 1948, 8:454-57.
17. Memorandum by Magill to Division of Chinese Affairs, 26 January 1949, *FRUS*, 1948, 8:461-62.
18. Sidney Souers, "Memoirs," 15-16 December 1954, Post-Presidential Files, Truman Library.
19. Stuart to Marshall, 29 January 1948, *FRUS*, 1948, 8:465-67.
20. Ibid., p. 466.
21. "Legislative History of the China Aid Act," p. 2, *Congressional Record* 94, 80th Cong. 2d sess., 1948, pp. 1395-96.
22. Marshall to Stuart, 17 February 1948, *FRUS*, 8:474-75.
23. Dean G. Acheson, "Legislative-Executive Relations," in Acheson, *This Vast External Realm*, pp. 220-21.
24. It is interesting to note that the Senate passed the European Recovery Program before the China Aid Act came up for debate. It was not necessary to add a provision for China in the ERP in the Senate. This modifies somewhat the conclusions by Tsou and Feaver. See Tsou, *America's Failure in China*, p. 465; John H. Feaver, "The China Aid Bill of 1948: Limited Assistance as a Cold War Strategy," *Diplomatic History* 5 (Spring 1981):117.
25. "Memorandum of the Secretary" 19 February 1948, Office Files of Assistant Secretary of State for Economic Affairs, 1946-1947, Truman Library.
26. "Comparison of China Aid Bill with Foreign Aid Act," 19 February 1948, Office Files of Assistant Secretary of State for Economic Affairs, 1946-1947, Truman Library.
27. Tsou, *America's Failure in China*, p. 468.
28. "Legislative History of the China Aid Act," pp. 3-4.
29. Ibid., pp. 4-5.
30. Ibid., pp. 4-6.
31. *Congressional Record* 94, 80th Cong., 2d sess., 1948, p. 3668.
32. Ibid.
33. Ibid., pp. 3669-71.
34. Ibid., p. 3672.
35. Ibid., p. 3694.
36. Ibid., pp. 3678-85.
37. See ibid., pp. 3683-85.
38. This paragraph specified that "not less than 5 nor more than 10 percent" of the funds for economic assistance were to be allocated for a rural reconstruction program. The funds were to be supplied in either U.S. dollars, proceeds in Chinese currency from the sale of ECA commodities, or both. "Legislative History of the China Aid Act," p. 7.

39. Souers, "Memoirs."
40. Butterworth complained that the China Aid Program was more similar to "an expanded post-U[nited] N[ations] R[elief and] R[ehabilitation] A[dministration] relief program" than to the ERP. Butterworth to Marshall, 24 January 1948, *FRUS*, 1948, 8:460.
41. Marshall to Stuart, 17 February 1948, *FRUS*, 1948, 8:474-75.
42. See Lovett to Forrestal, 25 October 1948, *FRUS*, 1948, 8:672.
43. Stuart to Marshall, 21 October 1948, *FRUS*, 1948, 8:671.
44. Lovett to Forrestal, 25 October 1948, *FRUS*, 1948, 8:672-73.
45. It is also clear that there were other tensions between Forrestal and the administration. He was dismissed within a few months when Truman thought he "was going to crack up." Memorandum, 11 September 1950, Truman Papers, President's Secretary's File, Truman Library.
46. See, for example, Ross Y. Koen, *The China Lobby in American Politics* (New York: Macmillan, 1976).
47. Lovett to Marshall, 12 November 1948, *FRUS*, 1948, 8:201-202.
48. Tsiang to Marshall, 12 November 1948, *FRUS*, 1948, 8:677-80.
49. Lovett to Stuart, 12 November 1948, *FRUS*, 1948, 8:202-203.
50. Douglas to Marshall, 9 December 1948, *FRUS*, 1948, 8:683-84.
51. See, for example, Waldo Heinrichs, "Roosevelt and Truman: The Presidential Perspective," in *Uncertain Years: Chinese-American Relations, 1947-1950*, ed. Dorothy Borg and Waldo Heinrichs (New York: Columbia University Press, 1980), pp. 3-12.
52. Stuart to Marshall, 26 November 1948, *FRUS*, 1948, 8:297-98; Marshall to Stuart, 26 November 1948, ibid., pp. 298; Butterworth to Marshall, 2 December 1948, ibid., pp. 298-99; Memorandum of Conversation by Marshall, 3 December 1948, ibid., pp. 299-301.
53. Cabinet Meeting, 26 November 1948, Connelly Papers, Truman Library.
54. Memorandum of Conversation by Marshall, 3 December 1948, *FRUS*, 1948, 8:299-301.
55. Lovett to Stuart, 13 December 1948, *FRUS*, 1948, 8:302.
56. Memorandum of Conversation by Marshall, 27 December 1948, *FRUS*, 1948, 8:302-304.
57. Carter to Lovett, 27 December 1948, *FRUS*, 1948, 8:304-305.
58. Memorandum of Conversation by Marshall, 27 December 1948, *FRUS*, 1948, 8:302-304.
59. Memorandum of Conversation by Lovett, 27 December 1948, *FRUS*, 1948, 8:305-306.
60. Memorandum by the Joint Chiefs of Staff to Forrestal, 24 November 1948, *FRUS*, 1949, 9:261-62.
61. Lovett to Souers, 14 December 1948, *FRUS*, 1948, 8:339-42.

62. See Ambassador Philip D. Spouse, chief, Division of Chinese Affairs, 1948, Oral History Interview, Truman Library, p. 36.

63. Lapham to Hoffman, 26 November 1948, *FRUS*, 1948, 8:654-55.

64. Cleveland to Lapham, 21 December 1948, *FRUS*, 1948, 8:658-62.

65. Butterworth to Lovett, 30 December 1948, *FRUS*, 1948, 8:667-68.

66. Lovett to Krentz, 10 December 1948, *FRUS*, 1948, 8:227-28.

67. Edgar to Acheson, 6 May 1949, *FRUS*, 1949, 9:328; "The Ownerless Isle," *The Economist*, 21 May 1949. In June 1949 the Office of Far Eastern Affairs prepared a statement explaining the basis for partially revoking the Cairo Declaration concerning Taiwan in order to gain UN support for an independence movement. Butterworth to Rusk, 9 June 1949, *FRUS*, 1949, 9:346.

68. Butterworth to Lovett, 16 December 1948, *FRUS*, 1948, 8:234.

69. Stuart to Marshall, 18 December 1948, *FRUS*, 1948, 8:235-36.

70. Central Intelligence Agency, "Review of the World Situation," CIA 12-48, Truman Papers, President's Secretary's File, Truman Library.

71. Taiwan was not the only focus for the possible use of the United Nations. At this same time the State Department was pushing for UN responsibility over the fate of South Korea. On January 1, 1949, the United States extended de jure recognition to the new government of the Republic of Korea, and one of the goals for U.S.-Korea relations was the withdrawal of American troops there, since the southern part of the peninsula was "liberated" from Japan by the United States after World War II. Any future problems in the area were to be handled multilaterally by the United Nations and not solely by the United States. U.S. Department of State, "U.S. Policy Regarding Korea," part 3, December 1945-June 1950, Research Project no. 252, December 1955, Truman Library.

72. Lovett to Truman, 14 January 1949, *FRUS*, 1949, 9:265-67.

73. Ibid.

74. Butterworth to Rusk, 9 June 1949, *FRUS*, 1949, 9:346.

75. Lovett to Truman, 14 January 1949, *FRUS*, 1949, 9:267.

76. Memorandum prepared in the Office of Far Eastern Affairs, 14 January 1949, *FRUS*, 1949, 9:599-601.

77. Hoffman to Lovett, 19 January 1949, *FRUS*, 1949, 9:270; Memorandum of Conversation by Butterworth, undated, ibid., pp. 601-606.

78. Ibid., p. 602.

79. Memorandum prepared in the Office of Far Eastern Affairs, 14 January 1949, *FRUS*, 1949, 9:566-601; Butterworth to Acheson, 15 March 1949, ibid., p. 607.

80. Lapham to Hoffman, 14 March 1949, Sumner Papers, Truman Library; Lapham to Sumner, "State Department Position on China Legislation," 22 March 1949, Sumner Papers, Truman Library.

81. Acheson to Connelly, 15 March 1949, *FRUS*, 1949, 9:607-609.

82. Tsou, *America's Failure in China*, p. 500.
83. Acheson to Connelly, 15 March 1949, *FRUS*, 1949, 9:607–609.
84. "Chinese Instructions to Ambassador Koo," 10 May 1949, Truman Papers, President's Secretary's File, Harry S. Truman Library.
85. This was not well received by the State Department. Acheson reported to the president at a cabinet meeting on May 13 that Chennault "is a better soldier than politician. He is misguided by his loyalty to the reigning families in the Nationalist government." Cabinet meeting, Friday the thirteenth, May 1949, Connelly Papers, Truman Library.
86. "Editorial Note," *FRUS*, 1949, 9:610.

Chapter 3

1. CIA, "Review of the World Situation," CIA 1-49, Truman Papers, President's Secretary's File, Truman Library.
2. Cabot to Acheson, 28 January 1949, *FRUS*, 1949, 8:91–92.
3. National Security Council (NSC), "A Report of the President by the National Security Council on the Current Position of the United States with Respect to Formosa," NSC 17/2, 3 February 1949, Truman Papers, President's Secretary's File, Truman Library.
4. Ibid.
5. Memorandum by Souers to the NSC, 4 February 1949, *FRUS*, 1949, 9:282.
6. Acheson to Stuart, 14 February 1949, *FRUS*, 1949, 9:287–88.
7. Note by Souers to the NSC, 18 February 1949, *FRUS*, 1949, 9:288–89.
8. Ambassador Livingston Merchant, Oral History Inverview, Truman Library.
9. Note by Souers to the NSC, 11 February 1949, *FRUS*, 1949, 9:284–86.
10. Note by Souers to the NSC, 1 March 1949, *FRUS*, 1949, 9:290–92.
11. Memorandum by Souers to the NSC, 3 March 1949, *FRUS*, 1949, 9:294–96.
12. Ibid., p. 296.
13. Sidney Souers, "Memoirs," 15–16 December 1954, Post-Presidential File, Truman Library.
14. Stuart to Acheson, 28 March 1949, *FRUS*, 1949, 8:207–208.
15. Acheson to Edgar, 30 March 1949, *FRUS*, 1949, 9:305–306.
16. Edgar to Acheson, 6 March 1949, *FRUS*, 1949, 9:297; Edgar to Acheson, 9 March 1949, ibid., p. 298; Stuart to Acheson, 31 March 1949, ibid., p. 306; Merchant to Butterworth, 24 May 1949, ibid., pp. 337–41.
17. Edgar to Acheson, 6 April 1949, *FRUS*, 1949, 9:308.
18. Acheson to Edgar, 15 April 1949, *FRUS*, 1949, 9:315–16.
19. Acheson to Souers, 18 May 1949, *FRUS*, 1949, 9:335–36.
20. Edgar to Acheson, 6 May 1949, *FRUS*, 1949, 9:328; Edgar to Ach-

eson, 8 June 1949, ibid., pp. 345–46; MacDonald to Acheson, 23 July 1949, ibid., pp. 364–65; Clark to Acheson, 18 May 1949, *FRUS*, 1949, 8:669–70.

21. Merchant, Oral History Interview.

22. Edgar to Acheson, 10 May 1949, *FRUS*, 1949, 9:330–31; Edgar to Acheson, 11 May 1949, ibid., p. 331.

23. Butterworth to Acheson, 21 April 1949, 893.01/4-49, National Archives, Washington, D.C.

24. Clark to Acheson, 18 May 1949, *FRUS*, 1949, 8:327–28; Clark to Acheson, 18 May 1949, ibid., pp. 669–70.

25. Peace talks ended after Li refused to meet certain Communist demands, such as turning over Chiang for trial as a war criminal. See Memorandum by Butterworth, 1 June 1949, *FRUS*, 1949, 9:701–705.

26. Clark to Acheson, 23 May 1949, *FRUS*, 1949, 8:340–41.

27. Stuart to Acheson, 24 May 1949, *FRUS*, 1949, 8:341–42.

28. Clark to Acheson, 6 August 1949, *FRUS*, 1949, 8:476–77.

29. Memorandum of Conversation by Butterworth, 1 June 1949, *FRUS*, 1949, 9:701–705.

30. Ibid., pp. 704–705.

Chapter 4

1. Mukden is another name for Shenyang. Since State Department records consistently refer to it as Mukden and not by its Chinese name, I will do so here also.

2. Clubb to Marshall, 26 November 1948, *FRUS*, 1948, 7:840; Memorandum of Telephone Conversation by Sprouse, 8 December 1948, ibid., p. 843; Lovett to Hopper, 19 December 1948, ibid., p. 844.

3. Stuart to Marshall, 16 December 1948, *FRUS*, 1948, 7:846.

4. Lovett to Stuart, 23 December 1948, *FRUS*, 1948, 7:849; Stuart to Marshall, 12 January 1949, *FRUS*, 1949, 8:934–35; Lovett to Stuart, 19 January 1949, *FRUS*, 1949, 8.

5. Memorandum of Conversation by Sprouse, 30 December 1948, *FRUS*, 1948, 7:703.

6. U.S. Department of State, OIR Report no. 5011, 28 July 1948, p. 56.

7. Ibid., p. iv.

8. Lovett to Stuart, 17 December 1948, *FRUS*, 1949, 7:659.

9. See Memorandum by Sprouse, 3 January 1949 and 6 January 1949, *FRUS*, 1949, 9:1–2, 5–6.

10. British Embassy to Department of State, 5 January 1949, *FRUS*, 1949, 9:4–5.

11. Memorandum of Conversation by Sprouse, 6 January 1949, *FRUS*, 1949, 9:5–6.

12. Acheson to Clubb, 2 March 1949, *FRUS*, 1949, 9:952.
13. British Embassy to Department of State, 19 March 1949, *FRUS*, 1949, 9:11-12.
14. Acheson to Clubb, 15 April 1949, *FRUS*, 1949, 9:952; Acheson to Paddock, 18 March 1949, ibid., p. 947; Stuart to Acheson, 6 April 1949, ibid., p. 951; Paddock to Acheson, 25 March 1949, ibid., p. 948; Hopper to Acheson, 4 March 1949, ibid., p. 948; Hopper to Acheson, 4 March 1949, ibid., p. 946; Clubb to Acheson, 26 March 1949, ibid., p. 949.
15. Cabot to Acheson, 30 April 1949, *FRUS*, 1949, 9:14; Cabot to Acheson, 27 April 1949, ibid., p. 1251; Acheson to Clubb, 26 April 1949, *FRUS*, 1949, 8:955; Clubb to Acheson, 30 April 1949, ibid., p. 955.
16. Stuart to Acheson, 3 May 1949, *FRUS*, 1949, 9:14-15.
17. Stuart to Acheson, 5 May 1949, *FRUS*, 1949, 9:16-17.
18. Acheson to Certain Diplomatic and Consular Offices, 6 May 1949, *FRUS*, 1949, 9:17.
19. Acheson to Stuart, 13 May 1949, *FRUS*, 1949, 9:21-23.
20. Stuart to Acheson, 7 June 1949, *FRUS*, 1949, 8:961-62.
21. Clubb to Acheson, 26 March 1949, *FRUS*, 1949, 8:949-50.
22. Stuart to Acheson, 26 March 1949, *FRUS*, 1949, 8:962; Clubb to Acheson, 9 June 1949, ibid., p. 963.
23. Clubb to Acheson, 1 June 1949, *FRUS*, 1949, 8:357-60.
24. Webb to Stuart, 4 June 1949, *FRUS*, 1949, 8:367; Clubb to Acheson, 2 June 1949, *FRUS*, 1949, 9:363-64; Stuart to Acheson, 6 June 1949, ibid., pp. 368-69; Stuart to Acheson, 7 June 1949, ibid., p. 372; Clark to Acheson, 6 June 1949, ibid., p. 370.
25. James E. Webb, "Meeting with the President," 16 June 1949, 893.00/6-1649, National Archives, Washington, D.C.
26. Clubb to Acheson, 11 June 1949, *FRUS*, 1949, 8:964.
27. Clubb to Acheson, 19 June 1949, *FRUS*, 1949, 8:965-67. For Stuart, this was yet another bit of evidence that the Soviets were instigating the Chinese Communists' actions in Manchuria. He noted, for example, the mention of alleged espionage activities in Inner Mongolia. Such activity was not directed against the Chinese people or the CCP, but toward the Soviets. See Stuart to Acheson, 21 June 1949, *FRUS*, 1949, 8:970.
28. Stuart to Acheson, 30 June 1949, *FRUS*, 1949, 8:766-67.
29. Davies to Kennan, 30 June 1949, *FRUS*, 1949, 8:768-69.
30. Cabot to Acheson, 1 July 1949, *FRUS*, 1949, 8:769; Ambassador John M. Cabot, Oral History Interview, Truman Library.
31. Acheson to Stuart, 1 July 1949, *FRUS*, 1949, 8:769.
32. Cabot, Oral History Interview. Cabot explains this last opinion with a description of what he considered to be the State Department's assessment of Ambassador Stuart. "Ambassador Stuart was a *dear* [emphasis in original] person. He knew China inside out, had many, very many good friends in China

on both sides of the fence, but he had no discretion; he simply couldn't keep a secret. He had a Chinese secretary, who knew everything he knew, and who was reporting right to Chiang Kai-shek. The result was that the Department didn't really trust the reports it was getting from the Embassy, and I was supposed to report by private letter to Walt Butterworth on what was going on.''

33. Ibid.

34. Cabot to Acheson, 7 July 1949, *FRUS*, 1949, 8:1202-1204; Cabot, Oral History Interview.

35. Cabot to Acheson, 8 July 1949, *FRUS*, 1949, 8:1207; Cabot to Acheson, 8 July 1949, ibid., pp. 1208-09; Stuart to Acheson, 8 July 1949, ibid., pp. 1211-12.

36. Cabot to Acheson, 9 July 1949, *FRUS*, 1949, 8:1220-22; Acheson to Cabot, 12 July 1949, ibid., p. 1225.

37. Stuart to Acheson, 28 June 1949, *FRUS*, 1949, 9:47; Memorandum of Conversation by the Assistant Chief of the Division of Western European Affairs, 7 July 1949, ibid., pp. 48-49; Foster to Acheson, 15 July 1949, ibid., p. 49; Douglas to Acheson, 17 August 1949, ibid., pp. 56-61. It should also be noted that Zhou Enlai allegedly expressed his desire for the State Department to send a copy of his message to the British. See Clubb to Acheson, 1 June 1949, *FRUS*, 1949, 8:357. Although it is not clear from available documents that his wish was carried out, developments in Sino-British relations at this time revealed that some contact was probably made.

38. Jones to Acheson, 19 August 1949, *FRUS*, 1949, 9:974.

39. Tozer, ''Last Bridge to China,'' p. 66; National Security Council, NSC 41, 28 February 1949, *FRUS*, 1949, 9:826-34.

40. National Security Council, NSC 41, pp. 826-34.

41. Memorandum by Souers to the Council, 3 March 1949, *FRUS*, 1949, 9:834.

42. Clark to Acheson, 6 June 1949, *FRUS*, 1949, 9:1098; Webb to Clark, 20 June 1949, ibid., p. 1099.

43. British Embassy to Department of State, 22 June 1949, *FRUS*, 1949, 9:1101.

44. Cabot to Acheson, 22 June 1949, *FRUS*, 1949, 9:1101; Acheson to Clark, 24 June 1949, pp. 1104-05; Clark to Acheson, 28 June 1949, ibid., p. 1114.

45. Butterworth to Acheson, 27 June 1949, *FRUS*, 1949, 9:1110-12.

46. British Embassy to Department of State, 4 August 1949, *FRUS*, 1949, 9:1125-26; British Embassy to Department of State, 20 July 1949, ibid., p. 1122; Memorandum of Telephone Conversation by Sprouse, 13 August 1949, ibid., p. 1130; British Embassy to Department of State, 29 July 1949, ibid., pp. 1124-25; Hopper to Acheson, 11 August 1949, 893.00/8-1149, National Archives, Washington, D.C.

47. Douglas to Acheson, 2 July 1949, *FRUS*, 1949, 9:866-67.

48. Acheson to Douglas, 29 July 1949, *FRUS*, 1949, 9:867-68.
49. Acheson to Souers, 14 April 1949, *FRUS*, 1949, 9:842-44; Hawthorne to Acheson, 2 August 1949, ibid., p. 1021.
50. Acheson to Douglas, 24 June 1949, *FRUS*, 1949, 9:1016-17; Acheson to Douglas, 30 June 1949, ibid., pp. 1018-19; Memorandum by Merchant 19 August 1949, ibid., pp. 1022-24.
51. Memorandum by Merchant, 19 August 1949, *FRUS*, 1949, 9:1022-24.
52. Gross to Knowland, 23 August 1949, *FRUS*, 1949, 9:1025-26.
53. Memorandum by Sprouse to Merchant, 26 August 1949, *FRUS*, 1949, 9:1026-27.
54. An indication of Acheson's feelings toward Knowland and the Senator's accusations was revealed when Acheson was asked in 1953 if he felt Knowland was sincere in his charges. Acheson commented, "The question of sincerity of the United States Senator is beyond me. What I do know, and I am convinced, is that he is an honest man, he is an intense fellow, he gets all worked up, the blood rushes to his head, he takes on a kind of a wild, stary look at you, and I just do not think his mind works in a normal way when he gets excited. He is a violent partisan, a terrific fighter; and the word 'sincerity' just ceases to have any meaning. He is not wrapped up in his cause; but he goes in swinging. I think that decribes it better than any other way I know." Acheson, Princeton Seminars.

Chapter 5

1. U.S. Department of State, *United States Relations with China* (Washington, D.C.: Office of Public Affairs, 1949), p. iii.
2. Acheson, Princeton Seminars.
3. *United States Relations with China*, p. xvi.
4. Sprouse, Oral History Interview.
5. Acheson to Truman, 12 May 1949, *FRUS*, 1949, 9:1365-67. See, for example, Clifford to Acheson, 17 May 1949, ibid., p. 1367; Memorandum of Conversation by Webb, 13 June 1949, ibid., p. 1368; "Memo," 14 July 1949, Elsey Papers, Truman Library.
6. Acheson, Princeton Seminars.
7. Ibid.
8. "Memo," undated, Elsey Papers, Truman Library; Memorandum of Conversation by Webb, 13 June 1949, *FRUS*, 1949, 9:1368; Memorandum for the President by Clifford, 6 July 1949, ibid., pp. 1370-72; Johnson to Acheson, 21 July 1949, ibid., pp. 1377-85.
9. Acheson, Princeton Seminars.
10. Butterworth to Acheson, 15 July 1949,, *FRUS*, 1949, 9:1373-74; Acheson, Princeton Seminars; Johnson to Acheson, 21 July 1949, *FRUS*, 1949, 9:1381; Clark to Acheson, 11 July 1949, ibid., p. 1373.

11. Memorandum by Acheson of a Conversation with Truman, 18 July 1949, *FRUS*, 1949, 9:1374.

12. Memorandum by Acheson of a Conversation with Truman, 21 July 1949, *FRUS*, 1949, 9:1377; Memorandum by Acheson of a Conversation with Truman, 25 July 1949, *FRUS*, 1949, 9:1385-86.

13. Acheson to Truman, 29 July 1949, *FRUS*, 1949, 9:1388-90; Acheson to Truman, 3 August 1949, ibid., pp. 1391-92.

14. Acheson, Princeton Seminars.

15. Sprouse, Oral History Interview.

16. Tsou, *America's Failure in China*, pp. 509-11. Tsou agrees with Sprouse on the reason for the paper's early release. Also see *Congressional Record* 95, 82nd Cong., 1st sess. 1949, pp. 10813, 10875, 10941; Dean Acheson, Memorandum of Conversation with Truman, 18 August 1949, Acheson Papers, Truman Library.

17. MacDonald to Acheson, 6 August 1949, *FRUS*, 1949, 9:1392.

18. Clark to Acheson, 10 August 1949, *FRUS*, 1949, 9:1393; MacDonald to Acheson, 15 August 1949, ibid., p. 1395.

19. Jones to Acheson, 19 August 1949, *FRUS*, 1949, 9:1339-40.

20. Stuart to Acheson, 6 June 1949, *FRUS*, 1949, 8:750-51.

21. Memorandum by the Department of State to Souers, 4 August 1949, *FRUS*, 1949, 9:369-70.

22. Note by Souers, 22 August 1949, *FRUS*, 1949, 9:376-78.

23. MacDonald to Acheson, 28 August 1949, *FRUS*, 1949, 9:378; MacDonald to Acheson, 30 August 1949, ibid., pp. 380-81; Memorandum of Conversation by Freeman, 9 September 1949, ibid., pp. 388-89.

24. Note by Souers, 6 October 1949, *FRUS*, 1949, 9:392-97.

25. Acheson to MacDonald, 18 November 1949, *FRUS*, 1949, 9:428-31.

26. Truman to Li, 10 October 1949, *FRUS*, 1949, 8:550.

27. Strong to Acheson, 19 October 1949, *FRUS*, 1949, 8:554; Clark to Acheson, 16 August 1949, ibid., p. 493; Lutkins to Acheson, 8 February 1949, 893.01 Yunnan/2-849, National Archives, Washington, D.C.

28. Lutkins to Acheson, 15 November 1949, *FRUS*, 1949, 8:594-95.

29. Strong to Acheson, 23 October 1949, *FRUS*, 1949, 8:560-61; Lutkins to Acheson, 24 November 1949, ibid., p. 585.

30. Acheson to Strong, 26 October 1949, *FRUS*, 1949, 8:567; Strong to Acheson, 10 November 1949, ibid., pp. 582-83; *United States Relations with China*, p. xvii.

31. Strong to Acheson, 10 November 1949, *FRUS*, 1949, 8:582-83.

32. Senate Foreign Relations Committee Supplemental Notes on Executive Session, 12 October 1949, Truman Papers, President's Secretary's File, Truman Library; CIA, "Limitations of South China as an Anti-Communist Base," ORE 30-48, Truman Papers, President's Secretary's File, Truman Library.

33. Strong to Acheson, 28 September 1949, *FRUS*, 1949, 8:540-41; Lut-

kins to Acheson, 28 October 1949, ibid., p. 569.

34. Lutkins to Acheson, 9 November 1949, *FRUS*, 1949, 8:582; Similar messages were sent April, May, and July. See Webb to Lutkins, 22 November 1949, ibid., p. 605.

35. Lutkins to Acheson, 6 November 1949, *FRUS*, 1949, 8:621-22; Webb to Lutkins, 22 November 1949, ibid., p. 605; Acheson to Lutkins, 30 November 1949, ibid., p. 616.

36. Acheson to Clubb, 15 September 1949, *FRUS*, 1949, 8:978; Clubb to Acheson, 23 September 1949, ibid., pp. 979-80; Webb to Clubb, 4 October 1949, ibid., p. 980; Clubb to Acheson, 2 October 1949, *FRUS*, 1949, 9:103; Clubb to Acheson, 1 October 1949, *FRUS*, 1949, 8:544-45.

37. Document Transmitted by French Embassy to Department of State, 6 October 1949, *FRUS*, 1949, 9:103.

38. "Meeting with the President," 17 October 1949, 893.01/10-1749, National Archives, Washington, D.C.

39. Clubb to Acheson, 10 October 1949, *FRUS*, 1949, 9:117-18.

40. See, for example, Webb to Donovan, 7 October 1949, *FRUS*, 1949, 9:110; Doolittle to Acheson, 8 October 1949, ibid., p. 112; Caffrey to Acheson 9 October 1949, ibid., p. 115; Acheson to Certain Diplomatic and Consular Officers, 12 October 1949, ibid., pp. 122-23.

41. Ward to Clubb, 12 October 1949, *FRUS*, 1949, 8:981; Clubb to Acheson, 26 October 1949, ibid., p. 984; Clubb to Acheson, 28 October 1949, ibid., p. 987; Acheson to Clubb, 28 October 1949, ibid., pp. 988-89.

42. Clubb to Acheson, 29 October 1949, *FRUS*, 1949, 8:989-92; Clubb to Acheson, 3 November 1949, ibid., pp. 999-1000; Clubb to Acheson, 4 November 1949, ibid., pp. 1000-10002.

43. Memorandum by the Acting Secretary to State, 31 October 1949, *FRUS*, 1949, 9:355.

44. Butterworth to Webb, 4 November, *FRUS*, 1949, 8:1002-1003; Clubb to Acheson, 4 November 1949, ibid., p. 1004; Webb to Certain Diplomatic and Consular Officers, 9 November 1949, ibid., p. 1005; Acheson to Certain Diplomatic Representatives, 18 November 1949, ibid., pp. 1009-10.

45. Clubb to Acheson, 23 November 1949, *FRUS*, 1949, 8:1020; Clubb to Acheson, 28 November 1949, ibid., pp. 1024-26.

46. Club to Acheson, 28 November 1949, *FRUS*, 1949, 8:1024-26; Clubb to Acheson, 26 November 1949, ibid., p. 1021; Clubb to Acheson, 27 November 1949, ibid., p. 1023.

47. Acheson to Clubb, 1 December 1949, *FRUS*, 1949, 8:1032.

48. Acheson to Clubb, 1 December 1949, *FRUS*, 1949, 8:1033; Wellborn to Acheson, 10 December 1949, ibid., pp. 1044-46.

49. See, for example, Dunn to Acheson, 7 December 1949, *FRUS*, 1949, 9:217; Vincent to Acheson, 1 December 1949, ibid., pp. 209-10; McConaughy to Acheson, 21 December 1949, ibid., p. 235; CIA, "Review of the

World Situation," January 1950, CIA 1-50, Truman Library.

50. Memorandum of Conversation by Sprouse, 24 December 1949, *FRUS*, 1949, 9:242. The British had waited, under pressure from the State Department, until the General Assembly session which convened in September 1949 had ended in order to avoid a debate on Chinese representation in the United Nations at this early date. See Memorandum of Conversation by Jessup, 10 November 1949, *FRUS*, 1949, 2:205. This will be discussed in detail in the next chapter.

51. Cabinet Meeting, 22 December 1949, Connelly Papers, Truman Library.

52. McFall to Acheson, 29 November 1949, 893.01/11-749, National Archives, Washington, D.C.; Meeting with the President, 20 December 1949, Acheson Papers, Truman Library.

Chapter 6

1. Acheson to Strong, 12 September 1949, *FRUS*, 1949, 9:1133.

2. Memorandum of Conversation by the Acting Secretary of State, 1 October 1949, *FRUS*, 1949, 9:1141; Memorandum of Conversation by Acheson, 19 October 1949, ibid., p. 1147; Memorandum by Acheson to Truman, 20 October 1949, ibid., pp. 1150-52.

3. Acheson to McConaughy, 20 October 1949, *FRUS*, 1949, 9:1152; Acheson to Strong, 20 October 1949, ibid., pp. 155-56.

4. McConaughy to Acheson, 29 October 1949, *FRUS*, 1949, 9:1156.

5. McConaughy to Acheson, 30 October 1949, *FRUS*, 1949, 9:1157. John M. Cabot, consul general at Shanghai froom 1948 to 1949, has noted that American businessmen in China had other reasons in addition to the profit motive to defy the Nationalists. Cabot claimed that the American business community hated the Nationalists because they had taken away many privileges from foreign business since the end of World War II. The Nationalists were insisting on "equal treaties" with the foreign communities and the Americans "just couldn't get used to the idea that Chinese were sovereign there and equal with Americans as individuals." Cabot, Oral History Interview.

6. Webb to Truman, 31 October 1949, *FRUS*, 1949, 9:1157-58.

7. British Embassy to Department of State, 1 November 1949, *FRUS*, 1949, 9:151-54.

8. Bacon to Acheson, 8 November 1949, 893.01/11-849 EB, no. 2389, National Archives, Washington, D.C.

9. British Embassy to Department of State, 8 November 1949, *FRUS*, 1949, 9:186-87.

10. Neal to O'Sullivan and Sprouse, 4 November 1949, 893.00/1-449, National Archives, Washington, D.C.

11. Memorandum by Perkins to the Office of Chinese Affairs, 5 November 1949, *FRUS*, 1949, 9:168-72; MacDonald to Acheson, 4 November 1949, ibid., p. 1150; Acheson to Strong, 5 November 1949, ibid., p. 1160; MacDonald to Acheson, 7 November 1949, ibid., pp. 1160-61.

12. Meeting with the President, 7 November 1949, Acheson Papers, Truman Library; CIA, "Information Report," S03120, 7 November 1949, Truman Library; 893.01/11-1449, 14 November 1949, National Archives, Washington, D.C.

13. Note by Acting Executive Secretary to the NSC, 7 November 1949, *FRUS*, 1949, 9:889-90.

14. Acheson to Souers, 4 November 1949, *FRUS*, 1949, 9:890.

15. Memorandum by the Under Secretary of State, 14 November 1949, *FRUS*, 1949, 8:1008.

16. Memorandum by Acheson to Truman, 21 November 1949, *FRUS*, 1949, 8:1015.

17. Robert M. Blum has concluded that the consultants with whom Truman met were from the Office of Far Eastern Affairs. The original documents made no suggestion as to their identity. See Blum, *Drawing the Line: The Origin of the American Containment Policy in East Asia* (New York: Norton, 1982), p. 159.

18. Conversation with the President, 17 November 1949, Acheson Papers, Truman Library.

19. Memorandum of Conversation by Brown, 9 March 1950, *FRUS*, 1950, 6:622-23.

20. Memorandum by Bishop to Jessup, 6 June 1950, *FRUS*, 1950, 6:636-38; Acheson to Johnson, 28 April 1950, ibid., pp. 632-36. The second memorandum was also sent to Sawyer.

21. Interview with Robert Aylward, 30 March 1980, Washington, D.C.

22. 793.02/3-1450, 14 March 1950, Department of State, Washington, D.C.; 611.93/1-950, 9 January 1950, Department of State, Washington, D.C.; Memorandum by Webb to the President, 10 January 1950, *FRUS*, 1950, 6:270-72. It should be noted that the French did not follow the British lead in recognizing the PRC in January although they had promised to do so. After the outbreak of the Korean War, they became staunch supporters of American actions toward Korea, Taiwan, and the PRC because of a fear for the future of Indochina. France did not recognize the PRC until 1965.

23. Memorandum by Webb to the President, 10 January 1950, *FRUS*, 1950 6:270-72; Acheson to Clubb, 10 January 1950, ibid., p. 275.

24. Clubb to Acheson, 10 January 1950, *FRUS*, 1950, 6:273-75.

25. Interview with Robert Aylward.

26. U.S. Department of State, *Bulletin*, 13 January 1950, p. 119.

27. Interview with Robert Aylward.

28. CIA, "Chinese Nationalist Attack on US Shipping," 12 January 1950,

President's Secretary's File, Truman Library.

29. Quoted in Nancy Bernkopf Tucker, "Nationalist China's Decline," p. 162. See also Tozer, "Last Bridge to China," p. 74.

30. Sprouse to Merchant, 16 February 1950, *FRUS*, 1950, 6:312.

31. Acheson to Embassy in China, 8 February 1950, *FRUS*, 1950, 6:306–307.

32. CIA, "Chinese Nationalist Attack on US Shipping"; Department of State, "Blockade of China and the Hong Kong Aircraft Problem," 27 March 1950, Acheson Papers, Truman Library.

33. See Acheson to the Embassy in France, 11 February 1950, *FRUS*, 1950, 6:308–11; McConaughy to Acheson, 26 January 1950, ibid., pp. 296–300; McConaughy to Acheson, 21 January 1950, ibid., pp. 289–93; CIA, "Review of the World Situation," CIA 2-50, 15 February 1950, President's Secretary's File, Truman Library; 611.93/3-150, 1 March 1950, Department of State, Washington, D.C.

34. CIA, "Review of the World Situation," CIA 3-50, 15 March 1950, President's Secretary's File, Truman Library.

35. Rusk to Acheson, 14 April 1950, *FRUS*, 1950, 6:327–28.

36. 125.714/4-1250, 12 April 1950, *FRUS*, 1950, 6:329; Acheson to Sawyer, 8 June 1950, ibid., pp. 638–39; Memorandum to Telephone Conversation by Freemont, 29 June 1950, ibid., p. 640.

37. Acheson to Rankin, 12 July 1950, Department of State, Topical File, Truman Library; Perkins to Acheson, 13 July 1950, Department of State, Topical File, Truman Library; Rankin to Acheson, 7 July 1950, Department of State, Topical File, Truman Library.

38. Perkins to Acheson, 13 July 1950; Douglas to Acheson, 21 July 1950; Acheson to Douglas, 21 July 1950; Memorandum of Telephone Conversation to Jackson, 19 July 1950; and Douglas to Acheson, 21 July 1950, all found in Department of State, Topical File, Truman Library.

39. Letter to Averell Harriman, undated, Department of State, Topical File, Truman Library; Department of Commerce, OIT-665, 3 December 1950; OIT-664, 3 December 1950, Truman Library.

40. For a similar conclusion, see Tucker, "Nationalist China's Decline," p. 163.

41. Acheson, Princeton Seminars.

42. Sidney W. Souers, "Memoirs," 15–16 December 1954, Post-Presidential Files, Truman Library.

43. Warren I. Cohen, "Acheson and China," in *Uncertain Years*," ed. Borg and Heinrichs, p. 51.

44. OIR Report no. 5011, 28 July 1948, *FRUS*, 1948, 7:56.

45. Michael H. Hunt, "Mao Tse-tung and the Issue of Accommodation with the United States, 1948–1950," in *Uncertain Years*, ed. Borg and Heinrichs, pp. 185–234; Steven M. Goldstein, "Chinese Communist Policy Toward

the United States: Opportunities and Constraints, 1944-1950," in ibid., pp. 235-78.

46. Tozer, "Last Bridge to China," p. 77.
47. Acheson to Smyth, 2 December 1949, *FRUS*, 1949, 8:1033-34.
48. Interview with Robert Aylward; Note by the Acting Executive Secretary to the NSC, 7 November 1949, *FRUS*, 1949, 9:889-96.
49. Warren I. Cohen, "Consul General O. Edmund Clubb on the 'Inevitably' of Conflict Between the United States and the People's Republic of China, 1949-1950," *Diplomatic History* 5, 2 (Spring 1981):165-68.

Chapter 7

1. Cabinet Meeting, 6 January 1950, Papers of Matthew J. Connelly, Truman Library.
2. It should be noted that not all who read the speech shared this interpretation. The Chinese Communists, for example, commented in a February 1, 1950, editorial in *Renmin ribao* that Acheson had expanded American interests in Asia by including certain islands such as the Ryukyus which were claimed by the Chinese. Herbert B. Bix, "Japan and South Korea in America's Asian Policy," in *Without Parallel*, ed. Frank Baldwin (New York: Pantheon, 1974), pp. 184-85.

Also, one American who did not agree with the China bloc's interpretation was John J. Muccio, ambassador to Korea at the time of the speech. He claimed that the implications of Acheson's speech were not correctly presented to the American people. It was not a new position, but one taken three years earlier. The 1950 congressional campaign changed the impact of the speech because Republicans altered its significance by emphasizing the so-called excluded areas. What Acheson was saying, according to Muccio, was that "the U.S. unilaterally would *have* [emphasis in original] to fight any aggression" committed against the areas he mentioned. In those areas beyond his perimeter, it would be a problem for the United Nations with American aid. Ambassador John J. Muccio, Oral History Interview, Truman Library.

3. Rusk to Acheson, 30 May 1950, *FRUS*, 1950, 6:351.
4. Acheson to Butterworth, 27 December 1949, 893.01/12-2749 CSA, National Archives, Washington, D.C.
5. Sebald to Acheson, 22 June 1950, *FRUS*, 1950, 6:366-67.
6. See, for example, Clough, *Island China*, pp. 6-7; Tsou, *America's Failure in China*, p. 558; "Editorial Note," *FRUS*, 1950, 6:367.
7. Wellington Koo, "Notes of Conversation with Mr. Louis Johnson, Secretary of Defense, June 30, 1950 at the Pentagon," Koo Manuscript Collection, Columbia University.

8. Ambassador Edwin W. Martin, Oral History Interview, Truman Library.
9. British Embassy to the Secretary of State, 6 December 1949, Acheson Papers, Truman Library.
10. Memorandum of Conversation by Sprouse, 6 December 1949, *FRUS*, 1949, 9:435-36; Memorandum of Conversation by Acheson, 8 December 1949, ibid., pp. 442-43.
11. Acheson to Cowen, 19 December 1949, *FRUS*, 1949, 9:447.
12. Cowen to Acheson, 21 December 1949, *FRUS*, 1949, 9:450-51.
13. MacDonald to Acheson, 7 December 1949, *FRUS*, 1949, 9:441.
14. Acheson to Edgar, 19 December 1949, *FRUS*, 1949, 9:446-47.
15. Edgar to Acheson, 23 December 1949, *FRUS*, 1949, 9:454-55.
16. Koo to Acheson, 23 December 1949, *FRUS*, 1949, 9:457-60.
17. Butterworth to Acheson, 28 December 1949, *FRUS*, 1949, 9:461-63.
18. Secretary of Defense, "Possible United States Military Action Toward Taiwan Not Involving Major Military Forces," 27 December 1949, Truman Papers, President's Secretary's File, Truman Library; Note by Souers, 27 December 1949, *FRUS*, 1949, 9:460-61. The Joint Chiefs concluded by pointing out the connection between Communist moves toward Taiwan and those toward Indonesia. They stated that the "recommended action with respect to Formosa is a part of the overall problem of resisting the spread of Communist domination in East Asia. It is recognized that this is a piece-meal approach, as is [the] recommendation with respect to assistance to Indonesia forwarded . . . on 22 December 1949, but it is likewise a matter of urgency. These separate but related projects point up the necessity of early determination of an overall program for the solution of the major problem."
19. Memorandum of Conversation by Acheson, 29 December 1949, *FRUS*, 1949, 9:463-67.
20. Acheson to Edgar, 30 December 1949, *FRUS*, 1949, 9:468.
21. See Graves to Sprouse, 30 December 1949, *FRUS*, 1949, 9:468-69.
22. Memorandum, 5 January 1950, Acheson Papers, Truman Library.
23. "Meeting with the President," 5 January 1950, Acheson Papers, Truman Library.
24. Acheson to Johnson, 14 April 1950, *FRUS*, 1950, 6:325-26. It perhaps seems strange that Acheson would feel it necessary to explain the January statement to the secretary of defense in April. There appears to have been a conflict between the administration and Johnson, who had chosen to limit certain types of purchases by the Nationalists. This was not departmental policy. See Johnson to Acheson, 6 May 1950, *FRUS*, 1950, 6:339.

The following statement by Truman might explain why Johnson was not in touch with the State Department's policy toward Taiwan: "Today I had a most unpleasant duty to do. For some time I've known that Louis Johnson could not continue as Secretary of Defense. I've known him for thirty years personally—

helped make him National Commander of the American Legion, appointed him Treasurer of the National Democratic Committee in 1949, in which position he did a bang up job. Then I appointed him Secretary of Defense when I saw Jim Forrestal was going to crack up. Mr. Johnson did a good job with Forrestal and the unification of the armed forces. Then he came forth with a complex. I didn't understand what happened to him. He talked out of turn to the press, the Senate and the House—and kept it up. He talked to the lying, crazy columnists—Pearson, the Sop sisters (Alsops), Doris Fleeson, and others. He succeeded in making himself an issue both publicly and in Congress. He is inordinately ambitious and an egoist. When I told him that for the good of the country he'd have to quit, he said, 'you are ruining me.' That answers the question of what comes first—my sentimental attachment to Johnson or the country. I had to tell him he'd have to quit. I felt as if I'd whipped my daughter which I've never done. Johnson's done. Too bad. He had a grand opportunity." Memorandum, 11 September 1950, Truman Papers, President's Secretary's File, Truman Library.

25. Acheson to Embassy in China, 25 January 1950, *FRUS*, 1950, 6:293; Johnson to Acheson, 6 May 1950, ibid., p. 339; Freeman to Rusk, 14 June 1950, ibid., pp. 363–64.

26. ECA, China Aid Program, "Summary Statement and Program Requirements for the Period February 16, 1950-June 30, 1951," 19 January 1950, Papers of Clark M. Clifford, Truman Library.

27. Ibid.

28. Departments of State and Defense, "Immediate US Course of Action with Respect to Formosa," 3 August 1950, Truman Papers, President's Secretary's File, Truman Library.

29. Freeman to Rusk, 14 June 1950, *FRUS*, 1950, 6:363–64.

30. Webb to Embassy in China, 26 May 1950, *FRUS*, 1950, 6:344–46.

31. Ibid., p. 344.

32. Ibid.

33. NSC, 28 November 1950, Elsey Papers, Truman Library.

34. Souers has noted that the military and the State Department were in conflict over this issue. According to Souers, the Joint Chiefs of Staff wanted to keep American troops out of Korea because "without mobilization of the whole country" the United States could not assure control of South Korea. In April 1949 Secretary of Defense Louis Johnson said, "If these troops [in Korea] are not out in one month, the State Department has to pay for them." Souers, "Memoirs."

35. Ibid.

Chapter 8

1. Austin to Acheson, 20 November 1949, *FRUS*, 1949, 9:235–36.

2. United Nations, Official Records, Security Council, 29 December 1949, Lake Success, New York, p. 1.

3. Ibid., pp. 1-3.

4. Bacon to Merchant, 30 December 1949, *FRUS*, 1949, 9:257-59.

5. The composition of the Security Council in January 1950 was China, Cuba, Ecuador, Egypt, France, India, Norway, USSR, United Kingdom, United States, and Yugoslavia. Three members present in December 1949—Argentina, Canada, and the Ukraine—were replaced by India, Yugoslavia, and Ecuador.

6. Bacon to Merchant, 30 December 1949, *FRUS*, 1949, 9:257-59.

7. Ibid.

8. Acheson to Austin, 5 January 1950, *FRUS*, 1950, 2:186-87.

9. Austin to Acheson, 9 January 1950, *FRUS*, 1950, 2:187-89. Rule 20 of the Security Council rules of procedure allows that the president should not preside over the Council during consideration of a particular question with which he is directly concerned. Under such circumstances the representative of the member next in English alphabetical order should be asked to preside over consideration of that question.

10. The text of the resolution read: "*Having considered* the statement made by the Central People's Government of the Chinese People's Republic on January 1950 to the effect that it considers the presence in the United Nations Security Council of the representative of the Kuomintang [Nationalist] group to be illegal and insists on the exclusion of that representative from the Security Council, *Decides* not to recognize the credentials of the representative referred to in the statement by the Central People's Government of the Chinese People's Republic and to exclude him from the Security Council." United Nations, Official Records, Security Council, 10 January 1950, pp. 2-3.

11. Ibid., pp. 3-4. The USSR and Yugoslavia cast the two negative votes; India abstained.

12. Ibid., pp. 4-10.

13. Austin to Acheson, 11 January 1950, *FRUS*, 1950, 2:191-94; Acheson to Austin, 12 January 1950, ibid., pp. 194-95.

14. It could also be argued that such moves were attempts to accelerate American recognition. It will be demonstrated that the Soviet walkouts forced the secretary general and several delegations to pressure the United States to allow the seating of the PRC representatives.

15. CIA, "Review of the World Situation," CIA 2-50, 15 February 1950, Truman Papers, President's Secretary's File, Truman Library.

16. United Nations, Official Records, Security Council, 12 January 1950, p. 2.

17. Ibid., pp. 10-17.

18. Ibid., 13 January 1950, pp. 1-15.

19. Ibid., 17 January 1950, pp. 13-16.

20. Acheson to Sayre, 13 January 1950, *FRUS*, 1950, 2:196; Acheson to Sayre, 18 January 1950, ibid., pp. 197–98.
21. Clubb to Acheson, 20 January 1950, *FRUS*, 1950, 2:200.
22. Ibid.
23. Austin to Acheson, 20 January 1950, *FRUS*, 1950, 2:201–202.
24. Memorandum of Meeting in the office of the Secretary of State, 21 January 1950, *FRUS*, 1950, 2:205–107.
25. Ibid.
26. Gross to Acheson, 27 January 1950, *FRUS*, 1950, 2:210–14.
27. Acheson to the Embassy in France, 3 February 1950, *FRUS*, 1950, 2:219; United Nations, Official Records, Security Council, 7 February 1950, p. 33; Acheson to the Embassy in the United Kingdom, 7 February 1950, *FRUS*, 1950, 2:223–24.
28. Memorandum of Conversation by Ross, 25 February 1950, *FRUS*, 1950, 2:227–28.
29. Memorandum by Ross, 7 March 1950, *FRUS*, 1950, 2:233–37.
30. Gross to Acheson, 11 March 1950, *FRUS*, 1950, 2:238–42.
31. Ibid.
32. Acheson to Certain Diplomatic Missions and Consular Offices, 23 March 1950, *FRUS*, 19450, 2:244. The U.S. position was summarized as follows:

> 1. US Mission to UN reports US position on question of Chi[nese] representation in UN not clearly understood by all other Members UNSC. Circulation to UN Members by SYG Lie of legal memo on question recognition of Chi Commies and Chi representation in UN organs has added element of confusion. Lie's memo based on premise that linkage between question of representation in UN and question of recognition by Member Govts is unfortunate from practical standpoint of UN operations and erroneous in legal theory.
>
> 2. US position which has been stated in SC and in other UN organs may be summarized as follows:
>
>   a. US recognizes Nat Govt;
>
>   b. US opposes unseating of Nat reps in UN or seating of Commies;
>
>   c. US believes question of representation is procedural and can be decided by each organ of UN by necessary majority;
>
>   d. US will accept parliamentary decision made by each organ on this question;
>
>   e. US believes each Member shld decide for itself how it will vote on question in light of its own circumstances and interests and its appraisal of best interest of UN.
>
> 3. Although reasons why US opposes seating Commie rep are generally those which also lead us not to recognize Commie regime, we recog-

nize that other govts may not have same approach subj and must determine for them selves how they vote against unseating Natl rep and expression of our views as outlined preceding para inevitably affect thinking of other govts on this question, because of UN interest involved we wish refrain from any efforts influence others or from activities which might subject us to charge of bringing pressure on other members of representation question. In other words we do not wish to interfere in any way with normal parliamentary procedures including free discussions among membs and decisions freely arrived at.

4. Important to note that present situation in UN organs arises not from US attitude, but from Sov walkout due its unwillingness accept majority decision. Blame for present situation lies with USSR and not with US or any other Member which continues recognize Nat Govt.

5. If question arises you may discuss with For Off for purposes of clarification.

Acheson

33. Raynor to Perkins, 29 June 1950, *FRUS*, 1950, 2:245.

34. United Nations, Official Records, Security Council, 25 June 1950, p. 4.

35. The Soviet Union had vetoed the application of the Republic of Korea for U.N. membership in April 1949. For the United States, this demonstrated "the Soviet Union's continuing hostility toward the duly constituted, lawful Government of Korea." Muccio to Acheson, 27 April 1940, Department of State, Selected Documents Relating to the Korean War, Truman Library.

36. Yugoslavia cast the sole positive vote. Egypt, India, and Norway abstained. The USSR was absent. *Yearbook of the United Nations, 1950* (New York: Columbia University Press, 1951), pp. 221-22.

37. The temporary UN Commission on Korea was established in 1947 to oversee the electoral process in North and South Korea. In 1948 it was asked to work for friendly relations between the two areas, and the next year the General Assembly voted to continue its existence until unification was achieved.

38. See *FRUS*, 1950, 7:202.

39. *Yearbook of the United Nations*, 1950, p. 223.

40. Raynor to Perkins, 29 June 1950, *FRUS*, 1950, 2:245.

41. Bacon to Freeman, 29 June 1950, *FRUS*, 1950, 2:246-47. In this memorandum Bacon also made the observation that the placing of the U.S. 7th Fleet to protect Taiwan and Truman's "giving orders" to the Nationalists not to invade the mainland weakened the American insistence that China was a great power and entitled to a Security Council veto. "Logically, the President's statement would appear to pave the way for our unseating the Nationalist Government representative though not for our supporting the seating of the Chinese Communist representative. A vacancy on the Security Council would,

however, create legal complications."

42. Acheson to Austin, 3 July 1950, *FRUS*, 1950, 2:247.

43. Acheson to Embassy in India, 4 July 1950, *FRUS*, 1950, 2:247-48.

44. Kirk to Acheson, 9 July 1950, Department of State, Selected Documents Relating to the Korean War, Truman Library. Moscow was the site chosen for this transmittal to avoid premature leaks if discussions took place in New Delhi or Washington.

45. Ibid. The Indian ambassador mentioned that his government had started thinking along the lines of the proposed formula because of the "Formosa question." He indicated that the "linking of the Korean and Formosan questions were creating difficulties for [the] G[overnment] O[f] I[ndia] in its sincere desire [to] back U.N. efforts [in] Korea whole-heartedly."

In the opinion of the Indian ambassador to China, K. M. Pannikar, "the problem of Formosa represented [the] most dangerous world war threat. China by invoking Sino-Russian treaty could demand Soviet assistance against [the] U.S." Henderson to Acheson, 9 July 1950, Department of State, Selected Documents Relating to the Korean War, Truman Library.

46. Acheson, Memorandum of Conversation with the President, 10 July 1950, Department of State, Selected Documents Related to the Korean War, Truman Library.

47. Kirk to New Delhi, 17 July 1950, Department of State, Selected Documents Relating to the Korean War, Truman Library.

48. Kirk to Moscow, 13 July 1950, Department of State, Selected Documents Relating to the Korean War, Truman Library.

49. Acheson to Austin, 29 July 1950, *FRUS*, 1950, 2:249-50; Acheson to Austin, 31 July 1950, ibid., pp. 251-52; Rusk to Acheson, 1 August 1950, ibid., p. 254.

50. *Yearbook of the United Nations, 1950*, p. 425.

51. The USSR, Yugoslavia, India, United Kingdom, and Norway voted in favor. China, Cuba, Ecuador, France, and the United States voted against. Egypt abstained. Ibid.

52. Meeting with the President, 3 August 1950, *FRUS*, 1950, 2:256-57.

53. Acheson to Embassy in India, 4 July 1950, *FRUS*, 1950, 2:247-48; British Embassy to Department of State, 11 August 1950, ibid., pp. 256-62; Douglas to Acheson, 11 July 1950, Department of State, Selected Documents Relating to the Korean War, Truman Library.

54. Although Acheson did not spell out what he meant by this statement, he apparently was referring to continued American aid to the Nationalists to keep the island from falling to the Communists.

55. Meeting with the President, 3 August 1950, *FRUS*, 1950, 2:256-57.

56. V. K. Wellington Koo, "Notes of a Conversation with Mr. Livingston T. Merchant, Deputy Assistant Secretary of State, June 29, 1950," Wellington Koo Papers, Columbia University.

57. United Nations, Official Records, General Assembly, 19 September 1950, pp. 1–8.
58. Ibid.
59. "Formosa Question in the United Nations General Assembly," undated, Truman Papers, President's Secretary's File, Truman Library.
60. Ibid.
61. United Nations, Official Records, General Assembly, 19 September 1950, p. 15.
62. *Yearbook of the United Nations, 1950*, p. 429; "Formosa Question in the United Nations General Assembly."

Chapter 9

1. Sidney Souers, "Memoirs," 15–16 December 1954, Post-Presidential Files, Truman Library.
2. Stuart to Marshall, 6 November 1948, *FRUS*, 1948, 7:532–33.
3. Memorandum by the Joint Chiefs of Staff to Forrestal, 24 November 1948, *FRUS*, 1949, 9:261–62.
4. Ambassador Philip D. Sprouse, Oral History Interview, Truman Library.
5. Edgar to Acheson, 6 May 1949, *FRUS*, 1949, 9:328; Edgar to Acheson, 8 June 1949, ibid., pp. 345–46; MacDonald to Acheson, 23 July 1949, ibid., pp. 364–65; Clark to Acheson, 18 May 1949, *FRUS*, 1949, 8:689–90.
6. U.S. Department of State, *United States Relations with China*, p. xvi.
7. U.S. Department of State, *Bulletin*, "Crisis in Asia—An Examination of U.S. Policy," February 1950.
8. See Butterworth to Acheson, 7 December 1949, *FRUS*, 1949, 9:438–40; Freeman to Rusk, 14 June 1950, *FRUS*, 1950, 6:363–64.
9. U.S. Department of State, OIR Report no. 50111, 28 July 1948.
10. Souers, "Memoirs."
11. Acheson, Princeton Seminars.
12. Cabinet Meeting, 22 December 1949, Connelly Papers, Truman Library.
13. 611.93/1-950, 9 January 1950, Department of State, Washington, D.C.; Memorandum by Webb to the President, 10 January 1950, *FRUS*, 1950, 6:270–72; U.S. Department of State, *Bulletin*, 13 January 1950, p. 119.
14. See Tucker, "Nationalist China's Decline," and Tozer, "Last Bridge to China," p. 74.
15. British Embassy to Department of State, 19 March 1949, *FRUS*, 1949, 9:11–12.
16. U.S. Department of State, "Blockade of China and the Hong Kong Aircraft Problem," 27 March 1950, Acheson Papers, Truman Library.

17. Acheson to Sawyer, 8 June 1950, *FRUS*, 1950, 6:638–39.

18. Acheson to Certain Diplomatic Missions and Consular Offices, 23 March 1950, *FRUS*, 1950, 2:243–44.

19. Bacon to Freeman, 29 June 1950, *FRUS*, 1950, 2:246–47; Raynor to Perkins, 29 June 1950, ibid., p. 245; Austin to Acheson, 3 July 1950, ibid., p. 248.

20. Major General C. A. Willoughby, "Intelligence Aspects of the Far East Command," 7 June 1950, Department of State, Selected Documents Relating to the Korean War, Truman Library.

21. Ibid.

22. "Points Requiring Presidential Decision," undated, Elsey Papers, Truman Library.

23. "President Truman's Conversations with George M. Elsey," 26 June 1950, Elsey Papers, Truman Library.

Bibliography

Acheson, Dean G. Papers. Harry S. Truman Library, Independence, Missouri.
———. *Present at the Creation* (New York: Norton, 1969).
———. Princeton Seminars, July 22–23, 1953. Papers of Dean Acheson. Harry S. Truman Library, Independence, Missouri.
———. *This Vast External Realm* (New York: Norton, 1973).
Ayers, Eben A. Papers. Harry S. Truman Library, Independence, Missouri.
Aylward, Robert. Interview. March 30, 1980. Washington, D.C.
Baldwin, Frank, ed. *Without Parallel: The American-Korean Relationship Since 1945* (New York: Pantheon, 1974).
Bernstein, Barton J., and Allen J. Matusow. *The Truman Admdinistration: A Documentary History* (New York: Harper & Row, 1966).
Blum, Robert M. *Drawing the Line: The Origin of the American Containment Policy in East Asia* (New York: Norton, 1982).
Borg, Dorothy, and Waldo Heinrichs, eds. *Uncertain Years: Chinese-American Relations, 1947–1950* (New York: Columbia University Press, 1980).
Brookings Institution. *Major Problems of United States Foreign Policy, 1949–1950* (Washington, D.C.: Brookings Institution, 1949).
Butterworth, W. Walton. Draft, Oral History Interview. Harry S. Truman Library, Independence, Missouri.
Cabot, John M. Oral History Interview, Harry S. Truman Library, Independence, Missouri.
Clayton, Will L. Papers. Harry S. Truman Library, Independence, Missouri.
Clifford, Clark M. Papers. Harry S. Truman Library, Independence, Missouri.
Clough, Ralph M. *Island China* (Cambridge: Harvard University Press, 1978).
Clubb, O. Edmund. Oral History Interview. Harry S. Truman Library, Independence, Missouri.
Cohen, Warren I. "Consul General O. Edmund Clubb on the 'Inevitability' of Conflict Between the United States and the People's Republic of China," *Diplomatic History* 5, 2 (Spring 1981): 165–68.
Connelly, Matthew J. Oral History Interview. Harry S. Truman Library, Independence, Missouri.
———. Papers. Harry S. Truman Library, Independence, Missouri.

Central Intelligence Agency. Reports. Papers of Harry S. Truman. Harry S. Truman Library, Independence, Missouri.
Congressional Record. 80th–81st Congress, 1947–1949, vols. 93–96.
Council on Foreign Relations. *The United States in World Affairs, 1949–1949* (New York: Harper, 1949).
───────. *The United States in World Affairs, 1949* (New York: Harper, 1950).
───────. *The United States in World Affairs, 1950* (New York: Harper, 1951).
Draper, William H. Oral History Interview. Harry S. Truman Library, Independence, Missouri.
Economic Cooperation Administration. *Economic Aid to China Under the China Aid Act of 1948* (Washington D.C., Economic Cooperation Administration, 1949).
Elsey, George M. Oral History Interview, Harry S. Truman Library, Independence, Missouri.
───────. Papers. Harry S. Truman Library, Independence, Missouri.
Feaver, John H. "The China Aid Bill of 1948: Limited Assistance as a Cold War Strategy," *Diplomatic History* 5, 2 (Spring 1981): 107–20.
Griffin, R. Allen. Oral History Interview. Harry S. Truman Library, Independence, Missouri.
Henderson, Loy W. Oral History Interview. Harry S. Truman Library, Independence, Missouri.
Hickerson, John D. Oral History Interview. Harry S. Truman Library, Independence, Missouri.
Hinton, William. *Iron Oxen* (New York: Vintage, 1970).
Hoffman, Paul G. Oral History Interview. Harry S. Truman Library, Independence, Missouri.
───────. Papers. Harry S. Truman, Independence, Missouri.
Kalicki, J. H. *The Pattern of Sino-American Crises* (New York: Cambridge University Press, 1975).
Koen, Ross Y. *The China Lobby in American Politics* (New York: Macmillan, 1960).
Koo, V. K. Wellington, Papers. Columbia University Library. New York, New York.
Locke, Edwin A. Oral History Interview. Harry S. Truman Library, Independence, Missouri.
───────. Papers. Harry S. Truman Library, Independence, Missouri.
McLellan, David S. *Dean Acheson, The State Department Years* (New York: Dodd, Mead, 1976).
Martin, Edwin W. Oral History Interview. Harry S. Truman Library, Independence, Missouri.
May, Ernest R. *The Truman Administration and China, 1945–1949* (Philadelphia; J. B. Lippincott, 1975).
Melby, John F. Papers. Harry S. Truman Library, Independence, Missouri.

Merchant, Livingston. Oral History Interview. Harry S. Truman Library, Independence, Missouri.
Millis, Walter, ed. *The Forrestal Diaries* (New York: Viking, 1951).
Muccio, John J. Oral History Interview. Harry S. Truman Library, Independence, Missouri.
―――. Papers. Harry S. Truman Library, Independence, Missouri.
Murphy, Charles S. Papers. Harry S. Truman Library, Independence, Missouri.
National Security Council. Reports. Papers of Harry S. Truman. Harry S. Truman Library, Independence, Missouri.
Paddock, Paul. *China Diary* (Ames: Iowa State University Press, 1977).
Purifoy, Lewis McCarrol. *Harry Truman's China Policy: McCarthyism and the Diplomacy of Hysteria, 1947-1951* (New York: New Viewpoints, 1976).
Souers, Sidney. "Memoirs" December 15-16, 1954. Post-Presidential Files. Harry S. Truman Library, Independence, Missouri.
Spanier, John. *American Foreign Policy Since World War II* (New York: Praeger, 1977).
Sprouse, Philip D., Oral History Interview. Harry S. Truman Library, Independence, Missouri.
―――. Papers. Harry S. Truman Library, Independence, Missouri.
Stolper, Thomas E. *China, Taiwan, and the Offshore Islands* (Armonk, N.Y.: M. E. Sharpe, Inc., 1985).
Stone, I. F. *The Hidden History of the Korean War* (New York: Monthly Review, 1952).
Tozer, Warren W. "Last Bridge to China: The Shanghai Power Company, the Truman Administration and the Chinese Communists," *Diplomatic History* 1,1 (Winter 1977): 64-78.
Truman, Harry S. *Memoirs.* 2 vols. (Garden City, N.Y.: Doubleday, 1955-1956).
―――. Papers. Harry S. Truman Library, Independence, Missouri.
Truman, Margaret. *Harry S. Truman* (New York: Morrow, 1973).
Tsou, Tang. *America's Failure in China, 1941-1950* (Chicago: University of Chicago Press, 1963).
United Nations. Documents. (Microfilm collection) Library of the Fletcher School of Law and Diplomacy, Tufts University, Medford, Massachusetts.
―――. *Economic Survey of Asia and the Far East, 1950* (New York: UN Department of Economic Affairs, 1951).
―――. Official Records. Lake Success, New York, 1949-1950. Library of the Fletcher School of Law and Diplomacy, Tufts University, Medford, Massachusetts.
U.S. Department of Defense. Korean War File. Harry S. Truman Library, Independence, Missouri.
U.S. Department of State. *Bulletin.*

———. 893.00 and 893.01. China files. Department of State, Washington, D.C.

———. 893.00 and 894.01. China files. National Archives, Washington, D.C.

———. *Foreign Relations of the United States* (Washington, D.C.: Government Printing Office, 1961–). Vols. 7 and 8, 1948 (1973); 8 and 9, 1949 (1978 and 1974); 1, 2, 6, and 7, 1950 (1977, 1976, 1976, and 1976).

———. Korean War File. Harry S. Truman Library, Independence, Missouri.

———. Topical File. Harry S. Truman Library, Independence, Missouri.

———. *United States Relations with China* (Washington, D.C.: Office of Public Affairs, 1949).

Yearbook of the United Nations, 1950 (New York: Columbia University Press, 1951).

Index

Acheson, Dean, 5-6, 12, 42, 48, 54, 55, 72, 78, 137, 183n.; address to National Press Club, 127, 141, 150, 160, 167, 189n.; and aid to Yunnan, 95, and China Aid Act, 20-21; and China White Paper, 81-85; and conflict in UN, 144-45, 148-50, 152-54, 156-61, 171, 194-94n.; disagreement with Truman, 108-13; and Merchant mission, 48; and military aid to Taiwan, 48, 92, 131, 136, 195n.; and Mukden hostage situation, 62, 98, 99, 100, 101, 168; and Nationalist blockade, 104, 105; and Nationalist bombing of Shanghai, 74, 116; and tensions with British, 75, 76, 97, 103, 117, 118, 159, 170; and trade with China, 77, 112, 120, 121, 123
American interests in China, 5, 10, 35, 166, 175n.
Anchises, 75
Anglo-American conflicts, 5, 8, 9, 73-78, 120, 123, 129, 169-70, 171
Anglo-Iranian Oil Co., 76
Austin, Warren R., 145, 146, 156
Aylward, Robert, 113, 114-15

Bacon, Leonard L., 106, 107
Bacon, Ruth G., 143, 144, 156, 194-95n.
Barr, David G., 29
Bebler, Ales, 146, 147, 148, 151, 155
Bevin, Ernest, 60, 101, 105

Bishop, Max W., 111-12
Bohlen, Charles E., 20-21
Bradley, Omar, 135, 137
Bridges, Styles, 18, 27, 43
Butterworth, W. Walton, 34, 54, 70, 75, 99, 128, 135, 182n.; and aid to Taiwan, 40, 131; and China Aid Act, 15, 17-18, 19, 26; and China White Paper, 82, 83, 177n.; and recognition of China, 61

Cabot, John M., 17, 46, 64, 70, 71, 73, 181-82n., 186n.
California-Texas Oil Co. (Caltex), 76, 77, 112, 120, 121
Carter, Marshall S., 31
Central Intelligence Agency: assessment of PRC/USSR treaty, 118; and Nationalist bombings, 117; reviews situation on Taiwan, 37; and Soviet walkout at UN, 147; and use of South China as base, 95
Chen Cheng, 48, 51, 54, 86, 131
Chennault, Claire, 43, 179n.
Chiang Kai-shek, 4, 18, 29, 36, 46, 54, 115, 119; resignation as president, 7; resumes presidency, 131, 134; retreats to Taiwan, 53, 54, 88, 91; U.S. support for, 123
Chiang Kai-shek, Madame, 30-31
China Aid Act, 14-44, 164-65, 166; amendment to, 39-43; and China bloc, 14-15; development of, 6, 15-

203

20, 25; implementation of, 35; and Nationalist demands, 18, 31
China bloc, 14–15, 18, 19, 28, 127, 164, 167, 168
China White Paper (*U.S. Relations with China*), 7, 28, 160, 161; development of, 80–85; and Taiwan, 85–91, 128, 167
Chu Changwei, 93, 94
Clark, Lewis, 54, 61, 64, 68, 70, 74, 75
Clifford, Clark M., 83, 84, 86
Clubb, O. Edmund, 9, 61, 62, 66, 98, 99, 101, 114; meeting with CCP representative, 119; on recognition of China, 125
Connally, Tom, 25, 42
Cowen, Myron M., 130

Davies, John P., 69, 70
Denfeld, Louis, 48
Dennison, Robert L., 28
Douglas, Lewis W., 30
Draper, William H., 27
Dulles, John F., 127, 128

Economic Cooperation Administration (ECA), 23, 34, 113, 165; and amendment to China Aid Act, 40; report on aid to Taiwan, 138–39
Edgar, Donald D., 52, 131, 134–35
Elsey, George M., 173
European Recovery Program, 21–22, 165, 176n.

Ford, J. F., 59, 60, 62
Foreign Assistance Act of 1948, 25, 28, 34. *See also* China Aid Act
Forrestal, James, 28, 112, 177n.
Franks, Oliver S., 60, 117, 118, 170
Fugh, Philip, 67

Gan Jiehou, 55, 56
Graves, Hubert A., 60, 61, 130
Gross, Ernest A., 77, 146, 151, 153, 154, 155, 159

Hoffman, Paul G., 34, 39, 41
Hopper, George D., 62, 63

Huang Hua, 67, 69, 70, 72, 125

India: delegation to UN, 156–58, 160, 161, 171
Isbrandtsen Co., 105, 116

Japan, 10, 32, 37, 138, 166
Jessup, Philip C., and China White Paper, 82, 83, 84
Jiang Tingfu, 29, 143–48
Johnson, Louis, 84, 128, 137, 190–91n.
Joint Chiefs of Staff, 7–8, 190n.; and China White Paper, 83, 84; and South China as base, 96; and Taiwan, 32, 89–91, 131–36, 139, 141, 165
Joint Commission on Rural Reconstruction, 25, 52, 138, 139
Joint U.S. Military Advisory Group in China (JUSMAG), 29
Jones, John W., 73
Judd, Walter H., 18, 27, 43, 85

Kavanaugh, E. P., 76, 78
Kem, James P., 24, 25, 27
Kennan, George F., 12, 69
Keon, Michael, 67
Kirk, Alan G., 157, 158
Knowland, William F., 43, 77, 78, 126, 183n.
Koo, V. K. Wellington, 29, 43, 135, 160
Korean War, 7, 11–12, 121, 122, 128, 140, 154–55, 159, 162, 171, 172–73, 195n.
Krentz, Kenneth C., 35, 136

Lapham, Roger D., 27, 28, 34, 41
Li Zongren, 7, 50, 54, 55, 62, 74, 75, 81, 86, 88, 92, 131
Lie, Trygve, 11, 142, 145, 149–53, 154, 162, 170
Lodge, Henry C., 85
Lovett, Robert A., 20, 28, 29, 31, 34, 37, 128; diverts ships to Taiwan, 35; and Mukden hostage situation, 58; and recognition of China, 60; and strategic importance of Taiwan, 32, 38, 165
Lu Han, 95
Lutkins, LaRue R., 94, 96

MacArthur, Douglas, 126, 127, 128
MacDonald, John J., 91, 92
Malik, Yakov A., 143, 145-48, 151, 158, 159
Mao Zedong, 66, 67, 68, 70, 107, 115, 118, 125
Marshall, George C., 5, 17, 29, 30, 59; and China Aid Act, 6, 15, 18, 20, 26, 164, 165; and China White Paper, 83, 93
Martin, Edwin W., 128-29
May, Ernest, 4
McCarran, Patrick, 42, 43
McConaughy, Walter P., 13
Merchant, Livingston T., 116, 128, 130, 139, 160; mission to Taiwan, 45-56; and trade with China, 77
Meyrier, Jacques, 64
Morse, Wayne, 24, 27
Muccio, John J., 167-68, 189n.
Mukden hostage situation, 8, 58-60, 82, 97-102, 168-69

National Security Council, 10, 32, 36; and establishment of the PRC, 91; policy toward Taiwan, 136; trade policies, 73, 108-109
Nationalist Party, 4, 6-7, 19, 165
Nehru, Jawaharlal, 156-58

Olive, William M., 71-72, 109

People's Republic of China: American connection with, 3; establishment of, 4, 91; recognition question, 98, 107, 122, 168, 187n.
Pepper, Claude, 24, 27
Price, Byron, 149

Raynor, G. Hayden, 156
Ross, John C., 151, 152, 153
Rusk, Dean, 127, 128, 139, 149

Sawyer, Charles, 112, 119
Shanghai, 74-75, 78, 108, 116, 122, 168
Shanghai Power Co., 10, 73, 116, 123, 124

Shell Oil Co., 76, 120, 121
Smith, H. Alexander, 126
Soong, T. V., 52
Souers, Sidney W., 18, 19, 26, 49, 76, 77, 87, 109, 123, 128, 141, 191n.
Sprouse, Philip D., 42, 59, 60, 61, 62, 116, 130; and China White Paper, 65; and trade with China, 78
Standard Oil Co. of California, 76, 112, 116, 120, 121
Stevenson, Ralph, 64, 108
Stokes, William N., 100
Strong, Robert C., 94, 95, 116
Stuart, John L., 17, 36, 51, 68, 71, 87, 165, 181-82n.; assessment of Li Zongren, 50, 55; and China Aid Act, 18-19, 20, 26; invitation to Yanjing University, 69, 70, 125; and Mukden hostage situation, 58, 99; and recognition of China, 64

Taft, Robert A., 126
Taiwan: aid to, 131-34, 168; American connections with, 3; and China White Paper, 85-91, 128, 167; evacuation of American personnel, 140; retreat of Chiang to, 53, 54, 88, 91; strategic importance of, 7, 32-39, 89-91, 131-36, 142, 165, 167, 178n.; UN intervention in, 38, 141, 166, 168
Timberman, Thomas S., 27
Tong, Hollington, 28
Truman, Harry S., 92; and aid to China, 29, 31; and China White Paper, 81, 82, 84, 167; and Chinese representation in UN, 159; disagreements with Acheson, 108-13; domestic pressure on, 123; and Indian resolution in UN, 158; and Korean War, 155; and Mukden hostage situation, 99, 169; and recognition of China, 101, 102, 168; and spread of communism, 173; statement on Taiwan, 113, 124, 126, 150, 160, 167; suggests coastal blockade, 109-110; supports Nationalist blockade, 104-108; and tensions with British, 97, 105, 106, 159; and Zhou demarche, 68

Truman Doctrine, 19

United Nations, 10-11, 12, 170-71, 186n., 192n; challenge to Nationalists' credentials 144, 147; Indian mediation on behalf of PRC, 156-58, 160, 161, 171; issue of Chinese representation, 142-49, 193-94n.; possible intervention in Taiwan, 38, 141, 166, 168, 178n.; role of Trygvie Lie in, 149-53

U.S. Department of State, 4, 8, 12, 169; and China White Paper, 7; and evacuation of American personnel in Taiwan, 140; and problems in UN, 143, 146, 150, 153, 156, 160, 162-63; and strategic importance of Taiwan, 37-39, 142, 166

USSR, 4, 194n.; objectives in China, 13, 141, 168, 172, 181n.; treaty with PRC, 118; walkout of UN, 142, 148, 152, 192n.

Vandenberg, Arthur S., 15, 20-21, 23-24, 83

Vorys, John M., 22, 27
Vyshinsky, Andrey Y., 160

Walker, Melville H., 16
Ward, Angus T., 58, 68, 98, 99, 100, 107, 109, 124, 125
Webb, James E., 67, 68, 97, 99, 109, 111, 139; meetings with UN representatives, 149; report on British trade in China, 106
Willoughby, C. A., 172
Wood, Clinton T., 16
Wu, K. C., 46, 131, 132

Ye Gongzhao, 108
Yunnan, 93-96

Zheng Tianxi, 30
Zhou Enlai, 66, 67, 70, 97, 114, 115, 119, 122, 125, 182n.; communications to UN, 142, 143, 145, 149, 152; and Mukden hostage situation, 98, 101, 104
Zhu De, 97
Zinchenko, Konstantin, 151

About the Author

A graduate of Wellesley College, June M. Grasso received a Ph.D. in History from Tufts University in 1981. She has taught at Tufts and Bentley College and since 1984 has been an assistant professor of history at Boston University.

In addition to publishing several papers and reviews in scholarly journals, Professor Grasso is the co-editor (with John Zawaki) of *Readings in Chinese History* (1985).

East Gate Books

Harold R. Isaacs
RE-ENCOUNTERS IN CHINA

James D. Seymour
CHINA RIGHTS ANNALS 1

Thomas E. Stolper
CHINA, TAIWAN, AND THE OFFSHORE ISLANDS

William L. Parish, ed.
CHINESE RURAL DEVELOPMENT
The Great Transformation

Anita Chan, Stanley Rosen, and Jonathan Unger, eds.
ON SOCIALIST DEMOCRACY AND THE CHINESE LEGAL SYSTEM
The Li Yizhe Debates

Michael S. Duke, ed.
CONTEMPORARY CHINESE LITERATURE
An Anthology of Post-Mao Fiction and Poetry

Michiko N. Wilson
THE MARGINAL WORLD OF ŌE KENZABURO
A Study in Themes and Techniques

Thomas B. Gold
STATE AND SOCIETY IN THE TAIWAN MIRACLE

Carol Lee Hamrin and Timothy Cheek, eds.
CHINA'S ESTABLISHMENT INTELLECTUALS

John P. Burns and Stanley Rosen, eds.
POLICY CONFLICTS IN POST-MAO CHINA
A Documentary Survey, with Analysis

Victor D. Lippit
THE ECONOMIC DEVELOPMENT OF CHINA

James D. Seymour
CHINA'S SATELLITE PARTIES

June M. Grasso
TRUMAN'S TWO-CHINA POLICY